Bernhard Eduard Fernow

Forest Policies and Forest Management in Germany and

British India

Bernhard Eduard Fernow

Forest Policies and Forest Management in Germany and British India

ISBN/EAN: 9783337734824

Printed in Europe, USA, Canada, Australia, Japan

Cover: Foto ©berggeist007 / pixelio.de

More available books at **www.hansebooks.com**

FOREST POLICIES AND FOREST MANAGEMENT

IN

GERMANY AND BRITISH INDIA.

By B. E. FERNOW, LL. D.,

DIRECTOR NEW YORK STATE COLLEGE OF FORESTRY, CORNELL UNIVERSITY

Reprinted from H. Doc. No. 181, 55th Cong., 3d Sess.

WASHINGTON:
GOVERNMENT PRINTING OFFICE.
1899.

D. FOREST POLICIES OF EUROPEAN NATIONS.

The conditions which a hundred years ago influenced the policies of the European nations—namely, the necessity of looking out for continuance of domestic supplies—are at present well overcome, provided the supplies in other countries last and can readily be secured.

In regard to supplies, the European countries may be grouped into those which produce as yet more than they need, namely: Russia, Austria-Hungary, Servia, Sweden and Norway, which are, therefore, exporters; those which produce large quantities of forest products, but not sufficient for their needs, Germany, France, Switzerland; those which depend largely or almost entirely on importation, England, Belgium, Holland, Denmark, Spain, Portugal, Italy, Greece, and Turkey.

Nevertheless, at least in Germany, the desirability of fostering home production and advantages of a general economic character, especially employment of labor in winter time which the forest industries insure, have still an influence upon the policy of the Government, even with supply forests.

In this way may be explained the protective tariff against wood imports, which was enacted in 1885 and increased later, especially to keep out competition from the virgin woods of Austria-Hungary and Russia. The last revision of 1892 has for its object not the discouragement of importation, but the inducing of importation of only raw material to be manufactured at home, by imposing a duty five times as high on lumber as on logs.

The result, however, has been more satisfactory from the revenue point of view than in protecting the forest owners, the Austro-Hungarian railroads equalizing the duty charges by lower rates.

The existence of a State forest policy, such as most European States have adopted, is based at present mainly on the protective value of the forest cover and the recognition that private interest can not be expected, or is insufficient, to give proper regard to this feature in its treatment of the forest areas.

It can not be said that a finally settled policy exists in any of the States, not even in Germany, but only that it is in a highly advanced stage of formation, with the tendency of increasing governmental activity and interference.

Such a policy is expressed in various ways. State ownership, State supervision of communal and private forests, restriction of clearing and enforced reforesting, establishment of forestry schools, and experiment stations.

State ownership of forest areas, which in the beginning of the century began to decrease under the influence and misapplication of Adam Smith's teaching, and the doctrine of individual rights urged to its extreme consequences, is now on the increase in most States. Thus France, which during and after the Revolution, took the lead in this dismemberment of the forest property which the monarchy had maintained, sold during the years 1791 to 1795 nearly one-half of the State forests and continued to reduce the area until there remained in 1874 but one-fifth of the original holdings. Since then a reversal of the policy has been in practice, the area not only being increased but financial assistance in reforesting on a large scale being given to private owners and communities.

Thus in the budget for 1895 of $2,500,000 appropriated for the State forest department, $1,000,000 is set aside for the extension of the State forests and necessary improvement of the existing ones. The State owns about 2,600,000 acres—somewhat over 10 per cent of the total

205

area. In addition the private property is controlled entirely as regards clearing; that is to say, no clearing may be done without notice to the Government authorities, or, in the mountain districts, without sanction of the same.

This control is especially stringent with reference to the holdings of village and city corporations, which represent over 27 per cent of the forest area. These must submit their plans of management to the State forest department for approval, and are debarred from dividing their property, thus insuring continuity of ownership and conservative management.

The necessity for such control became apparent in the first quarter of the century, when as a consequence of reckless denudation in the Alps, Cevennes, and Pyrenees, whole communities became impoverished by the torrents which destroyed and silted over the fertile lands at the foot of the mountains. Some 8,000,000 acres of mountain forest in twenty departments were involved in these disastrous consequences of forest destruction, with 1,000,000 acres of once fertile soil made useless. The work of recovery was begun under laws of 1860 and 1864, and a revised law, the reboisement act, of 1882. Under this law the State buys and recuperates the land, or else forces communities or private owners to do so with financial aid from the Government.

Since the operation of this law the State has spent in purchases of worn out lands and in works to check the torrents and in reforesting, nearly $12,000,000, not including subventions to communities and private owners. It is estimated that $28,000,000 more will have to be expended before the area which the State does or is to possess, some 800,000 acres in all, will be restored.

A forestry school at Nancy educates the officers, and is among the best on the Continent.

England, in the home country, has had little need of a forest policy on account of its insular position and topography. Of the 3,000,000 acres of woodlands, mostly devoted to purposes of the chase or parks, 2 per cent are State forests, and so encumbered with rights of adjoining commoners as pasture or for wood supplies that no rational management is possible. But in India there is a well-organized forest administration with a very extensive area, namely, 60,000 square miles reserved and 34,590 square miles protected and under active control of the Government. The organization of the forestry service was begun in 1865 by German foresters. (See pages 259–263.) At present special schools of forestry, one in England and one in India, supply the technical education of the officers.

Italy has long suffered from the effects of forest devastation by droughts and floods, but the Government was always too weak to secure effective remedies. The State owns only 1.6 per cent of 116,000 acres of forest, the balance of 7,000,000 acres belonging to communities and corporations or individuals. Yet by the laws of 1877, revised in 1888, the policy of State interference is clearly defined. Excellent though the law appears on paper, it has probably not yielded any significant results or even general enforcement, owing to the financial disability of the Government. This law placed nearly half the area not owned by the State under Government control, namely, all woods and lands cleared of wood on the summits and slopes of the mountains above the upper limit of chestnut growth, and those that from their character and situation may, in consequence of being cleared or tilled, give rise to landslips, caving, or gullying, avalanches and snowslides, and may to the public injury interfere with water courses or change the character of the soil or injure local hygienic conditions. Government aid is to be extended where reforestation appeared necessary.

Of the 76,000 acres which required immediate reforestation, for reasons of public safety, only 22,000 were reforested in twenty years up to 1886, the Government contributing $85,000 toward the cost.

In the revised law of 1888, as a result of the vast experiences preceding, a further elaboration of the same plan was attempted by creating further authority to enforce action. It is now estimated that 531,000 acres need reforesting at a cost of $12,000,000, of which two-fifths is to be contributed by the State.

Expropriation proceedings may be instituted where owners refuse to reforest, with permission to reclaim in five years by paying the cost of work, with interest, incurred by the State.

In Austria, the disastrous consequences which the reckless devastation and abuse of her mountain forests by their owners has brought upon whole communities have led to a more stringent and general supervision of private and communal forests than anywhere else. Since 1885 there has been also in progress a work of recuperation similar to the French reboisement work, in which, up to 1894, nearly $1,500,000 had been spent, the State contributing variously from 25 to 100 per cent toward covering the expense. A fully organized forest department manages the Government forests, 2,000,000 acres, which are gradually being increased by purchase, or 73 per cent of the whole forest area. One higher, and several lower schools provide instruction.

Some 150,000 acres of waste land were reforested by the State between 1881 and 1890.

Even Russia, although one of the export countries, with $30,000,000 to $35,000,000, and largely in the pioneering stage, has a well-devised forest policy, developed within the last thirty or fifty years, which consists not only in maintaining Government forests to the extent of about 280,000,000 acres under tolerably good management, and 30,000,000 of Crown forests, personal property of the royal family, but in restricting private owners from abuse of their property, where the public welfare demands, while in the prairie country in southern Russia large amounts of money are spent by the Government in planting forests and assisting private enterprise in the same direction.

With the Siberian forests and those of the Caucasus added, the area of Government forest may reach the large figure of 600,000,000 acres, which, though not yet all placed under management, is sooner or later to come under the existing forest administration.

The restrictive policy dates from a very elaborate law passed in 1888, in which the democratic spirit in the constitution of the body controlling the exercise of property rights is interesting. The approval of working plans or of clearings on private property is placed in the hands of a specially constituted committee for each county, which includes the governor, justices of the peace, the county council, and several forest owners, and the Government itself must secure the approval of this committee for its operations.

By this law, throughout European Russia, woodlands may be declared "preserved forests" on the following grounds: That they serve as preventives against the formation of barrens and shifting sands, and the encroachment of dunes along seashores or the banks of navigable rivers, canals, and artificial reservoirs; that they protect from sand drifts towns, villages, cultivated land, roads, and the like; that they protect the banks of navigable rivers and canals from land-slides, overflows, or injuries by the breaking up or passing of ice; when growing on hills, steep places, or declines, they serve to check land or rock slides, avalanches, and sudden freshets, and all forests that protect the springs and sources of the rivers and their tributaries.

In these preserved forests, working plans are made at the expense of the Government, and in the unpreserved forests at the expense of the owners. In each province the Government maintains an inspector-instructor, whose duty is to advise those who apply to him in forest matters, and as far as possible he is to superintend on the spot all forestry work. The Government has established nurseries from which private owners can obtain young trees and seeds at a low price. The owners are allowed to employ as managers of their forests the trained officials of the forest administration, while medals and prizes are given yearly to private owners for excellency in forest culture and management. Two higher and thirteen lower schools of forestry are also maintained by the Government.

The country which has attracted most interest in all matters pertaining to forestry, because the science of forestry is there most developed and most closely applied, is Germany. The policies prevailing and methods employed are fully described in another part of this report.

It may, however, be interesting to trace somewhat the historical development both of the application of forestry principles and of the existing forest policy.

Although as early as Charlemagne's time a conception of the value of a forest as a piece of property was well recognized by that monarch himself, and crude prescriptions as to the proper use of the same are extant, a general really well-ordered system of forest management hardly existed until the beginning of the eighteenth century. Sporadically, to be sure, systematic care and regular methods of reproduction were employed even in the thirteenth and fourteenth centuries.

To understand the development of the present forest policy in Germany one must study the peculiar conditions and development of property rights that led to it. Germany was originally settled by warriors, who had to keep together in order to resist enemies and conquerors on every side, ready to move and change domicile at any moment. The soil which was conquered, consequently, was not divided, but owned as a whole, managed by and for the whole tribe. It is only in the sixth century that signs of private property in woodlands are discernible. Before that time it was *res nullius*, or, as it is expressed in legal manuscripts, *"quia non res possessa sed de ligno agitur."*

Wood being plentiful and yet needed by everybody, it appeared a crime only to take wood which had been already appropriated or bore unmistakable signs of ownership, such as being cut or shaped. But severe punishments were in earliest times inflicted for incendiarism and for damage to mast trees, since the seed mast for the fattening of swine was one of the most important uses of the forest.

There was not much need of partition, especially of the forests. The community, to which all the land of a district belonged, and which was managed by and for the aggregate of society, was called the "mark," a communistic institution of most express character, and every "marker" or shareholder was allowed to get the timber needed by him for his own use without control.

This early communal ownership of forest land undoubtedly explains the fact that even to-day over 5 per cent of the forest is owned by communities, cities, or villages. Gradually the necessity of regulating the cutting of the wood became apparent, as the best timber in the neighborhood of the villages was removed, and we find quite early mention of officials whose duty it was to superintend the felling, removing, and even the use of the timber. By and by even the firewood was designated by officials. Manufacturers received their material free of charge, but only as much as was needed to supply the community. Occasionally there were rules that each man had to plant trees in proportion to his consumption. So that by the end of the fourteenth century quite a system of forest management had been developed.

Meanwhile the Roman doctrine of the regal right to the chase had also begun to assert itself by the declaration of certain districts as ban forests or simply forests, in which the King exclusively reserved the right to chase. The Kings again invested their trusted followers and nobles with this right to the chase in various districts, thus gradually dividing the control of the same.

While at first these reservations did not bring with them restrictions in the use of the timber or pasture or other products of the forest, gradually these uses were construed as exercised only by permission, and the former owners were reduced to holders of "servitudes," i. e., holders of certain rights in the substance of the forests. The fact that the feudal lords frequently became the obermarkers or burgomasters of the mark community lent color of right to these restrictions in the use of the property, besides the assertion that the needs of maintaining the chase required and entitled them to such control.

It is interesting to note that through all the changes of centuries, these so-called servitudes have lasted until our own times, much changed, to be sure, in character, and extending by new grants especially to churches, charitable institutions, cities, villages, and colonists. Such rights, to satisfy certain requirements from the substance of an adjoining forest, were then usually attached to the ownership of certain farms, and involved counter service of some sort, usually in hauling wood or doing other forestry work.

Sometimes when the lordly owners of large properties exercised only certain prerogatives to show ownership, these, in the course of time, lapsed into the character of servitudes, the forest itself by occupation becoming the property of the community. With changes in value and other changes in economic conditions, these rights often became disadvantageous and more and more cumbersome to either or both sides.

The present century has been occupied with the difficult labor of relieving this state of things and making equitable arrangements by which the forests become unencumbered and the beneficiaries properly satisfied by cession of land or a money equivalent.

This chapter of the history of forest policy is especially interesting to us as a tendency, nay the practice exists of granting such rights to the public timber to the settlers in the Western States, which by and by will be just as difficult to eradicate when rational forest management is to be inaugurated.

Over 5,000,000 marks and several hundred acres of land were required in the little Kingdom of Saxony to get rid of the servitudes in the State forests. The Prussian budget contains still an item of 1,000,000 marks annually for this purpose; and although over 22,000,000 marks and nearly 20,000 acres of land have been spent for this purpose in Bavaria, the State forests there are still most heavily burdened with servitudes.

The doctrine of the regal right to the chase, as we have seen, led to the gradual assertion of all property rights to the forest itself, or at least to the exclusive control of its use. This right found expression in a legion of forest ordinances in the fifteenth and sixteenth centuries, which aimed at the conservation and improvement of forest areas, abounding in detailed technical precepts.

At first treating the private interest with some consideration, they gradually more and more restrict free management. Prohibition of absolute clearing, or at least only with the permission of the government; the command to reforest cleared and waste places; to foster the young growth; limiting the quality of timber to be felled: preventing devastation by prohibiting the pasturing of cattle in the young growth, of the removal of the forest litter, of pitch gathering, etc., were among these prescriptions, and many others, such as prescribing the manner and time of felling, the division into regular felling lots, determination as to what is to be cut as firewood and what as building timber. Then, with the increasing fear of a reduction in supplies, followed prohibitions against exportation, against sale of woodlands to foreigners, against speculation in timber by providing schedules of prices, and from time to time entire exclusion from sale of some valuable species. Even the consumer was restricted and controlled in the manner of using wood.

In mediaeval times, besides private forests of the King and lords, only the communal forest (allmende) was known, and small holdings of farmers were comparatively rare until the end of the Middle Ages.

The thirty years' war and the following troublesome times gave rise not only to extended forest devastation, but also to many changes in ownership of woodlands. With the growing instability of communal organization of the "mark," division of the common property took place, and thus private ownership by small farmers came about, reducing the communal holdings. Colonization schemes by holders of large estates also led to dismemberment.

A very large amount of the mark forest came into possesssion of the princes and noblemen by force, and later possessions of the princes were increased by the secularization of the property of monasteries and churches. Until the end of the last century these domains belonged to the family of the prince, just as the right to the throne or the governing of the little dukedom, contributing toward the expenses of government.

But when, as a consequence of the French Revolution and the Napoleonic wars and subsequent changes, the conception of the rights of the governing classes changed, and in some States like Prussia much earlier, a division of domains into those which belonged to the prince's family as private property and those which were State forests was effected, so that now the following classes of forest property may be distinguished:

(1) State forests, which are administered by the government for the benefit of the commonwealth, each State of the Confederation owning and administering its own.

(2) Imperial forests, belonging to and administered for the benefit of the Empire, situated in the newly acquired province of Alsace-Lorraine.

(3) Crown forests (Fidei-commiss), the ownership of which remains in the reigning family, administered by State government, but the revenues of which are in part applicable to government expenses.

(4) Princely domains, which are the exclusive and private property of the prince.

(5) Communal forests possessed by and administered by and for the benefit of village and city communities, or even provinces as a whole.

(6) Association forests, the owners of the old "mark" forests, possessed by a number of owners, the State sometimes being part owner.

(7) Institute and corporation, school or bequest forests, which belong to incorporated institutions, like churches, hospitals, and other charitable institutions.

(8) Private forests, of larger or smaller extent, the exclusive property of private owners.

The proportions of these classes of property which existed in the beginning of the century

H. Doc. 181——14

experienced considerable changes by the sale of State forests, the sales being due partly to financial distress, partly to a mistaken application of Adam Smith's theories, which supposed that free competition would lead to a better management and highest development of the forest industry as well as of other industries.

This tendency, however, was checked when the fallacy of the theory became apparent, especially with reference to a property that demands conservative treatment and involves such time element as we have seen.

The hopes which were based on the success of individualistic efforts were not realized, and although control of private action had been retained by the State authorities, this could not always be exercised, and the necessity of strengthening the State forest administration became apparent. The present tendency, therefore, is not only to maintain the State forests, but to extend their area by purchase, mostly of devastated or deforested areas and by exchange for agricultural lands from the public domain. Thus, in Prussia, the increase of State forest area has been at the rate of 14,000 acres per year since 1867.

In districts where small farmers own extensive areas of barrens a consolidation is effected, the parcels of remaining forest and the barrens are put together, the State acquires these and pays the owners either in money or other property.

In Prussia, during the decade 1882–1891, 30,000 acres were in this way exchanged for 17,000 acres, and in addition some 200,000 acres, waste or poorly wooded, purchased at an expense of $3,500,000, round numbers. During the same decade the reforestation of 80,000 acres of waste lands was effected, while nearly 75,000 acres in the State's possession remained to be reforested.

The annual budget for these reforestations of waste lands has been $500,000 for several years. The area of barrens and poor soils, only fit for forest purposes in Prussia, is estimated at over 6,000,000 acres.

The present distribution of the property classes for the whole Empire of the 35,000,000 acres of forest is about as follows, varying, to be sure, very considerably in the States of the Confederation:

	Per cent.
State and Crown forests (of which the Crown owns less than 2 per cent)	32.7
Imperial forests	1
Communal forests (5,000,000 acres)	15.2
Association forests	2.5
Institute forests	1.3
Private forests	18.3

The State and Crown forests are all under well-organized forest administrations, sometimes accredited to the minister of finance, sometimes to the minister of agriculture. These yield an annual net revenue of from $1 to $5 and $6 per acre of forest area, with a constant increase from year to year, which will presently be very greatly advanced when the expenditures for road building and other improvements cease.

In the State management the constant care is not to sacrifice the economic significance of the forest to the financial benefits that can be derived, and the amount cut is most conservative.

The Imperial forests are of course managed in the same spirit as the several State forests.

While the present communities, villages, towns, and cities are only political corporations, they still retain in some cases in part the character of the "mark," which was based upon the holding of property.

The supervision which the princes exercised in their capacity of Obermarker or as possessor of the right to the chase, remained, although based on other principles, as a function of the State when the "mark" communities collapsed, the principles being that the State was bound to protect the interest of the eternal juristical person of the community against the present trustees, that it had to guard against conflicts between the interest of the individual and that of the community in this property, and secure permanency of a piece of property which insured a continued and increasing revenue. The principle upon which the control of these communal holdings rests is then mainly a fiscal one.

The degree of control and restriction varies in different localities. Sale and partition and clearing can mostly take place only by permission of the State authorities, and is usually discountenanced except for good reasons (too much woods on agricultural soil).

With reference to 5.6 per cent of communal forest property, this is the only control which is of a fiscal nature. The rest is more or less closely influenced in the character of its management, either by control of its technicalities or else by direct management and administration on the part of the Government.

Technical control makes it necessary that the plans of management be submitted to the Government for sanction, and that proper officers or managers be employed who are inspected by Government foresters. This is the most general system, under which 49.4 per cent of communal forests are managed (as well in Austria and Switzerland), giving greatest latitude and yet securing conservative management. To facilitate the management of smaller areas several properties may be combined under one manager, or else a neighboring government or private forest manager may be employed to look after the technical management.

Where direct management by the State exists, the State performs the management by its own agents with only advisory power of the communal authorities, a system under which 45 per cent of the communal forests are managed (also in Austria and France).

In Prussia this system exists only in a few localities, but it is since 1876 provided as penalty for improper management or attempts to avoid the State control.

This system curtails, to be sure, communal liberty and possibly financial results to some extent, but it has proved itself the most satisfactory from the standpoint of conservative forest management and in the interest of present and future welfare of the communities. Its extension is planned both in Prussia and Bavaria.

Sometimes the State contributes toward the cost of the management on the ground that it is carried on in the interests of the whole commonwealth. A voluntary cooperation of the communities with the State in regard to forest protection by the State forest guards is in vogue in Wurttemberg, and also in France. Institute forests are usually under similar control as the communities.

The control of private forests is extremely varying. A direct State control of some kind is exercised over only 29.7 per cent of the private forest, or 14.6 per cent of the total area, mostly in southern and middle Germany, while 70.3 per cent of the private property, or 34.5 per cent of the total forest area, is entirely without control, a condition existing in Prussia and Saxony.

As far as the large land owners are concerned, this has mostly been of no detriment, as they are usually taking advantage of rational management; but the small peasant holdings show the bad effects of this liberty quite frequently in the devastated condition of the woods and waste places. As a recent writer puts it: "The freedom of private forest ownership has in Prussia led not only to forest dismemberment and devastation, but often to change of forest into field. On good soils the result is something permanently better; on medium and poor soils the result has been that agriculture, after the fertility stored up by the forest has been exhausted, has become unprofitable. These soils are now utterly ruined and must be reforested as waste lands.

Need, avarice, speculation, and penury were developed into forest destruction when in the beginning of this century the individualistic theories led to an abandonment of the control hitherto existing, and it was found out that the principle so salutary in agriculture and other industries was a fateful error in forestry.

Where control of private forests exists it takes various forms:

(1) Prohibition to clear permanently or at least necessity to ask permission exists in Wurttemberg, Baden, and partially in Bavaria. (Protection of adjoiners.)

(2) Enforced reforestation within a given time after removal of the old growth and occasionally on open ground where public safety requires.

(3) Prohibition of devastation or deterioration—a vague and undefinable provision.

(4) Definite prescription as to the manner of cutting (especially on sand dunes, river courses, etc.).

(5) Enforced employment of qualified personnel.

In addition to all these measures of restriction, control and police, and enforcement, there should be mentioned the measures of encouragement, which consist in the opportunity for the education of foresters, dissemination of information, and financial aid.

In the latter respect Prussia, in the decade 1882–1892, contributed for reforestation of waste places by private owners $335,000, besides large amounts of seeds and plants from its State nurseries. Instruction in forestry to farmers is given at twelve agricultural schools in Prussia. In nearly all States permission is given to Government officers for compensation, to undertake at the request of the owners the regulation or even management of private forest property.

For the education of the lower class of foresters there may be some twenty special schools in Germany and Austria, while for the higher classes not only ten special forest academies are available, but three universities and two polytechnic institutes have forestry faculties.

Besides, all States have lately inaugurated systems of forest experiment stations; and forestry associations, not of propagandists but of practitioners, abound. As a result of all this activity in forestry science and practice, not less than twenty forestry journals in the German language exist, besides many official and association reports and a most prolific book literature.

E. FOREST CONDITIONS AND METHODS OF FOREST MANAGEMENT IN GERMANY, WITH A BRIEF ACCOUNT OF FOREST MANAGEMENT IN BRITISH INDIA.

FOREST AREA, EXTENT AND OWNERSHIP.

Germany, as constituted at present, has an area of 133,000,000 acres—about one-fifteenth of our country—a population of about 47,000,000, or less than 3 acres per capita, or only one-tenth of our per capita average. Its forests cover 34,700,000 acres, or 26 per cent of the entire land surface. A large portion of the forests cover the poorer, chiefly sandy, soils of the North German plains, or occupy the rough, hilly, and steeper mountain lands of the numerous smaller mountain systems, and a small portion of the northern slopes of the Alps. They are distributed rather evenly over the entire Empire. Prussia, with 66 per cent of the entire land area, possesses 23.5 per cent of forest land, while the rest of the larger States have each over 30 per cent, except small, industrious Saxony, which lies intermediate, with 27 per cent of forest cover.

Considering the smaller districts of Prussia, Bavaria, and the smaller States, it is found that out of 64 provinces and districts, 18 have less than 20 per cent forest; 18 have from 20 to 29 per cent; 23, including the greater part of the country, have from 30 to 39 per cent, and 5 of the smaller districts have from 40 to 44 per cent of forest. The districts containing less than 20 per cent of forests are, as might be supposed, mostly fertile farming districts in which the plow land forms over 40 per cent of the land, but they also include neglected districts like Hanover and Luneburg, where a former shortsighted, selfish, and improvident policy has led to the deforestation of poor, flat lands, which have gradually been transformed into heaths, where an accumulation of bog-iron ore, and other obstacles render the attempts at reforestation difficult, expensive, and unsatisfactory. Left to forests, these same lands, which now are unable to furnish support to farmers or to produce a revenue to their owner, could easily pay the taxes and interest on a capital of $50 to $100 per acre. To reforest them now costs $10 to $50 per acre and requires a lifetime before any returns can be expected.

Since it is one of the common claims in the eastern United States that the land is all needed for agriculture, and since it will be conceded that in hardly any State east of the Mississippi much land necessarily remains untilled, it may be of interest to note that in this densely populated Empire of Germany out of 67 districts and provinces the plow land forms less than 20 per cent in 4 districts, 30 to 39 per cent in 10 districts, 40 to 49 per cent in 26 districts, 50 to 59 per cent in 20 districts, and 60 to 69 per cent in 7 districts, in spite of the fact that a large part of the forests are in private hands and would be cleared if the owners saw fit to do so.

In our country the total area in farms is only 18 per cent at present.

Of the total of 34,700,000 acres of forest land (an area about as large as the State of Wisconsin) 32.7 per cent belongs to the several States as State property; 19 per cent belongs to villages, towns, and other corporations, and 50 per cent to private owners, a considerable part of this being in large estates of the nobility.

213

The following figures show these ownership relations for the eight larger States, which involve 96 per cent of the total area of the empire:

State.	Population.	Total land surface.	Forests.				
			Total.	Per cent.	Owned by the		
					State.	Corpora tions.	Private.
	Millions.	M acres.	M acres.		M acres.	M acres.	M acres.
Germany	47	133,392	34,750	100	11,360	6,710	16,680
Prussia	29.9	88,000	20,240	58	6,190	3,240	10,900
Bavaria	5.6	18,200	6,200	18	2,160	890	3,150
Wurttemberg	1.9	4,800	1,470	4.2	480	470	530
Saxony	3.2	3,700	1,020	3	430	60	530
Baden	1.6	3,730	1,300	4	207	667	447
Alsace-Lorraine	1.5	3,600	1,100	3.1	360	520	220
Hesse	.9	1,900	500	1.7	170	220	200
Mecklenburg-Schwerin	.5	3,290	560	1.6	255	85	220

This same relation, expressed in per cent, becomes:

State.	Forest cover of total area	Forests owned by		
		States.	Corpora tions.	Private.
	Per cent.	Per cent.	Per cent.	Per cent.
Germany	25.7	32.7	19	48.3
Prussia	23.5	30	17	53
Bavaria	35	34	14	52
Wurttemberg	31	32	32	36
Saxony	27	43	6	51
Baden	37	18	49	33
Alsace-Lorraine	30	33	47	20
Hesse	31	29	37	34
Mecklenburg-Schwerin	17	46	15	39

The condition of the forests to a great extent depends on the degree of supervision or control exercised by the State authorities. It is best in all cases in the State forests, is equally good in the corporation forests under State control, and is poorest in the private forests, particularly those of small holders.

STATE CONTROL.

The amount of State influence or control varies in the several States, and varies in some cases even in one and the same State for different districts. Of the State forests, without exception, it can be said that they are nearly in that form which, according to present knowledge and with reasonable effort, is able to produce the greatest quantities of wood material in those dimensions and of such kinds as best to satisfy the demands of the markets and at the same time render the management as profitable as possible. This does not mean that they are not improving, for as forestry knowledge increases and the methods are perfected the results are better. From what follows it also appears that all State forests as a whole pay, and pay handsomely, when the low intrinsic value of the land on which the forest stocks is considered.

The control of the corporation forests is perfect only in a few of the smaller States, notably Baden, Hesse, and Alsace-Lorraine; also in some districts in Prussia where the corporation forests are managed by the State authorities, the wishes of the villagers or corporate owners being, however, always duly considered. In a large portion of Prussia, in Wurttemberg, and in Bavaria the corporation provides its own foresters; but these must be approved, as well as their plans of operations, by the State authorities, so that here the management is under strict control of the State, and favorable forest conditions at least partially assured. In Wurttemberg the corporation is given the choice of supplying its own foresters or else joining their forests to those of the State. This has led to State management of near 70 per cent of all corporation forests. Only the corporation forests of Saxony and those of a small part of Prussia are without any supervision. Of the private forests, those of Prussia and Saxony, involving 60 per cent of all private forests of the Empire, are entirely free from interference. They can be managed as the owner sees fit, and there is no obstacle to their devastation or entire clearing and conversion into field or pasture. The remainder of the private forests are under more or less supervision. In most districts a State permit is required before land can be cleared. Devastation is an offense, and in some States, notably Wurttemberg, a badly neglected forest property may be reforested and managed by State authorities. In nearly all States laws exist with regard to so-called "protective forests" i. e., forests needed to prevent floods, sand blowing, land and snow slides, or to insure regularity of water supply, etc. Forests proved to fall under this category are under special control, but as it is not easy in most cases to prove the protective importance of a forest, the laws are difficult to apply and rarely enforced.

A partial return to the State supervision of private forests has been attempted in Prussia by the establishment of a law which renders the owner of a forest liable for the damage which the devastation or clearing of his forest property causes to his neighbor. This law, however, like the former, is so difficult to apply, and puts the plaintiff to great expense, so that so far it has not been enforced to any extent except where the Government itself is the injured party.

In the following statement the areas of forest are grouped according to the degree of State supervision and manner of management:

(1) Managed by State authorities as State property, 11,360,000 acres, which is 32.7 per cent.

(2) Managed by the State authorities, but the property of corporations, villages, towns, etc., a little over 2,212,000 acres, which is 6.3 per cent.

(3) Under strict Government control, the plans of management and the permissible cut having to be approved by State authorities (corporation property), 3,875,000 acres, which is 11.1 per cent.

(4) Under supervision of the State, not only as common property but as special property, subject to inspection and, in part, to control of State forest authorities; nearly all private property and partly belonging to large estates, 4,767,000 acres, which is 13.7 per cent.

(5) Without any Government control or supervision beyond that of common property. These forests may be divided, sold, cleared, and mismanaged, except under the certain cases before mentioned. Here belong all private forests of Saxony and Prussia and part of the corporation forests of Prussia and all those of Saxony, 11,490,000 acres, which is 33 per cent.

CHARACTER OF FOREST GROWTH.

The greater part of the German forests is stocked with conifers, chiefly pine (the Scotch pine, a pine similar to our red or Norway pine) and spruce. The pine prevails on the sandy areas of North Germany, and occupies about 60 per cent of the Prussian and 30 per cent of the Bavarian forests. The spruce is the chief conifer and principal timber tree of Saxony and southern Germany. The hard woods, chiefly beech, some oaks, with small amounts of ash, maple, elm, etc., are most abundant in the valley of the Rhine, Lorraine, and Wurttemberg, but good beech forests occur in nearly all parts of the Empire.

The greater part of all forests of Germany are "timber forests," where the trees are cut at an age of over 80 years (generally 90 to 120 years).[1] Timber forests form over 90 per cent of the State forests of all larger States, are the prevalent form in the forests of corporations, and are common in those of private owners. The other two common forms, the "coppice" and "standard coppice," where the trees are cut at an age of less than 30 years (usually 15 to 25 years, and in the standard coppice a small part only is allowed to reach better age and size), are most abundant in private forests and to a less extent in corporation properties, but form only a very small part of the State woods, where they are steadily diminishing in importance. The coppice is a hard-wood forest, depends on the sprouting capacity of the trees, and furnishes small poles, firewood, and tanbark. Both forms of the coppice and standard coppice require a smaller amount of standing timber, furnish quicker returns, but do not furnish those kinds of products which the market demands in largest quantity.

In the timber forest the trees of any particular tract or division are supposed to be of about the same age, differing not over 20 years in the extreme, so that for a rotation of one hundred years, i. e., a management where the crop is harvested at the age of 100 years, one-fifth, or 20 per cent, of all the forests should be 1 to 20 years old; another 20 per cent, 21 to 40 years old, etc. In spite of the great difficulty of attaining this regularity of distribution in the forests of an entire State without disturbing the yearly cut of timber, this regularity is already attained very closely in most of the State forests. Thus in the State forests of Prussia, of the total area of

[1] For fuller description of the systems of management, see pp. 225 to 259 of this report.

timber forest (90 per cent of all State forests), the age of the timber is as follows: On 13 per cent of the area, over 100 years old: on 13 per cent, 81 to 100 years old; on 14 per cent, 61 to 80 years old; on 18 per cent, 41 to 60 years old; on 19 per cent, 21 to 40 years old; on 19 per cent, 1 to 20 years old, and about 4 per cent are clearings, where the timber has been cut lately.   In all forests the ground is at once reforested, if cut clean, or else the cut is so arranged that a natural seeding goes on as the harvest progresses, this latter consisting of several fellings, separated by a number of years.

EXPLOITATION.

The cutting in all State forests is generally done by the cord or by the cubic foot (really by the stere, festmeter, or cubic meter).   In rare cases the timber is cut and moved by the purchaser; nearly always it is cut and moved by the forest authorities and sold and delivered at the main roads.   The logs are not cut to uniform lengths, but care is had in the forest to cut to best advantage.   Long, straight timbers are left long, if possible, and sold as long, round, or sometimes hewn pieces; saw timber is cut in even lengths; poles are cut to suit local markets; wagon and coopers' stock, etc., are cut to suit, or left in round timbers, while pulp wood, cord wood, and branches, and sometimes even stumps, are worked up in customary manner, graded, and sold by the cord (really "stere" or "raummeter").

In the conversion of the logs into lumber there are more complications in dimensions than with us.   The measure is generally the meter and centimeter; edging is not done by even numbers. Lumber is sold by cubic measure, and the handling is thus generally not so simple as in America.

As far as practical means and methods in felling and logging operations go we can learn but little from Germany, except that more care in the utilization of the timber would be profitable here as it is abroad.   Yet it may be of interest, and not entirely devoid of suggestive value, to briefly recite the practices followed in most Government forests.

The location of fellings for the year having been determined with due consideration, the rangers engage and control, under supervision of the district manager, the crew of wood choppers under a foreman, who are mostly men living in the neighborhood of the range or district and accustomed to all kinds of forest work.[1]   A contract, which contains conditions, regulations, and a scale of prices, is made with them, which they sign.   The men are paid by the job, the prices per unit differing, of course, in different localities and being graded according to the kinds of timber, size, etc.

To cite one example we may take the schedule prices paid at the forest belonging to the city of Goslar, as this will interest us further on.   There are 40 men nearly permanently employed either in wood chopping, planting, or otherwise, and their average earnings during three years have been about 80 cents per working day.   The prices for cutting spruce, including moving to roads and barking, and the average prices obtained for ten years were as follows:

Cost of cutting.	Average price obtained in the woods.	
	Lowest class.	Highest class.
Saw timber, above 5 inches in diameter (5 classes), 85 cents per 100 cubic feet.	$9.50	$6.20
Long poles (3 classes), from $4 cents to $1.08 per 100 cubic feet.	5.90	7.10
Small poles (4 classes), from $1.37 to $3.07 per 100 cubic feet.	3.60	5.80
Firewood, split, 76 cents to $1 per cord.	3.60	4.20
Firewood, brush, $1.10 per cord.		1.60

In Prussia the average cost of lumbering (wood cutting and bringing to roads) for all kinds and dimensions is 65 cents per 100 cubic feet; that is to say, the wood-choppers' bill on the 300,000,000 solid cubic feet of wood harvested annually in the Prussian Government forests amounts to $1,950,000.   It will appear from the prices for wood cited that often the harvesting is more expensive than the price obtained, as, for instance, for brushwood, which will hardly sell for half the cost of cutting, but its removal is necessary from cultural considerations.   The wood choppers are also sometimes expected to move the cordwood at least to the neighboring roads, so as to obviate the driving of teams through the woods or young growth.

[1] In the census of Germany for 1881–82 there were reported as engaged in forestry, hunting, and fishing 384,637 persons.   Unfortunately, no division of the three occupations was made.

If the felling is to be a clearing, a strip is assigned to each gang of 3 men, 1 with an ax and 2 with saws (felling with the saw, of course, is the rule); if a regeneration cutting or thinning, the trees to be taken are carefully selected by the ranger or manager and marked with a marking hammer. As a rule, all fellings are done during winter, and all trees, except in the coppice and small poles, are felled with the saw close to the ground. In the pineries of the North German plain, where the root wood is salable, they are even dug out and then sawed off close to the root, thus saving a good piece of log timber, which in Saxony increases the wood value of the harvest by fully 3 per cent. Which parts of the log are to be cut into firewood and which into lumber wood or special timbers, and the length of the same according to the best use that can be made of the stick, are determined by the foreman, or in valuable timber by the ranger or manager himself. A scale of sizes and classes of timber (sortiment) exists; in general, all wood over 3 inches diameter is called Derbholz (coarse wood or lumber wood), all below 3 inches is brushwood (Reisholz), with which root wood (Stockholz) is classed. These last two grades are used as firewood, with which is also classed body wood or split wood (Scheitholz), split from pieces over 6 inches diameter at the small end, and round billet wood (Knüppelholz) of 3 to 6 inches diameter.

The wood to be used in the arts, called timber wood (Nutzholz), may appear either in bolts, corded, or in logs. The diameter measurement of logs is made by the ranger, with calipers, at the middle of the log. Every cord and every log is numbered and the diameter and length noted on the log, and a list prepared in which the cubic contents are calculated. From this list the manager checks off the result of the felling, marking each piece or cord with the marking hammer, and after advertisement sells at public auction, in the woods or at some public place, the single pieces or cords to the highest bidder over and above the Government rate, which for the different grades is established every three years on the basis of, but below current market prices. The sale of logs is made per cubic foot, and the size of the log influences the rate or price, heavier logs being disproportionately higher in price.

PRICE OF WOOD IN THE FOREST.

During the years 1881–1887 the following prices were obtained by the Prussian forest administration for wood in the forest. This is practically for stumpage, cut and marked, the buyer hauling it from the woods:

Price per 100 cubic feet of wood in Prussia.

Pieces containing 18–36 cubic feet.	Lowest price.	Highest price.	Average price.	
Timber.				
Oak	$8.50	$17.30	$12.00	14.00
Beech, ash, elm, maple	5.50	12.25	7.50	8.50
Spruce	4.75	11.65	7.00	8.00
Pine	4.75	11.00	6.25	6.35
Firewood:				
Beech, ash, elm, maple	.75	1.75	1.00	1.20
Spruce	.40	1.50	.70	.85
Pine	.45	1.30	.80	.90

To gain an idea of the appreciation of the wood product, without reference to kind, size, and quality, the following series of figures will serve:

Average price per 100 cubic feet of wood realized by the Prussian Government for its entire crop (about 500,000,000 cubic feet).

Year.	Price.
1850	$3.27
1855	3.66
1860	3.69
1865	4.71
1870	4.35
1875	5.21
1880	4.47
1885	4.30
1890	4.40

The highest price for any district was obtained in 1888, being $8.49, while the lowest was $2.82. The lower prices in later years are explained by the large importations of wood, especially from Hungary, Russia, and Sweden; for while our misinformed forestry friends point to Germany as the Eldorado of forestry and proclaim the proportion of forest area there maintained, namely, about 25 per cent, as the ideal and necessary for self-support, and therefore to be maintained also in this country, they overlook the fact that Germany imports not less than $60,000,000 worth of wood and wood manufactures, mostly of the same kind as grown or manufactured in that country. This represents about 10 per cent of the total consumption of Germany, while the importations of the United States, which imports from Canada only competing classes of forest products, represent not more than 1 per cent of our probable consumption.

The exports of forest products from Germany, on the other hand, are, to be sure, nearly 50 per cent of her imports, but they represent mostly manufactures, while in the United States the reverse is the case; that is to say, the United States exports twice as much as it imports, and that mostly raw material, namely, twice as much in value of raw material as of manufactures.

The countries from which Germany imports raw or partly manufactured wood are mainly Russia, Austria-Hungary, and Sweden, which furnish nearly five-sixths of the total importation, while Holland, England, Denmark, Belgium, France, and Switzerland draw about $14,000,000 worth of raw material from Germany. (See tables further on.)

To protect the forest owners of Germany, a tariff on importations was imposed in 1885 and increased later. Of the effects of this last measure a government report says that as a financial measure these tariffs have had excellent success, for the revenue from these duties increased from $646,000 in 1880 to $1,732,000 in 1886. But for the forest owner the hoped-for results did not become apparent; the Austro-Hungarian railroads and shipping interests lowered their rates so as to largely equalize the duty charges. The duties on unmanufactured materials being very low, the lack of results in the market of these is still more noticeable. Yet a salutary effect is stated to be a prevention of still lower prices, and because otherwise there would have been a lack of useful occupation for labor finding remunerative employment in the manufacture of the raw material, which, without the increase in duties, would have been imported in manufactured condition.

PRICE OF MANUFACTURED LUMBER.

The following samples of schedules for manufactured lumber, always delivered at the railroad station, may serve to give an idea to our lumbermen how nearly prices compare with those prevalent in our country. We choose those of eastern provinces, which are in sharpest competition with Russian and Hungarian imports:

Province of Posen.

Timber (7-8.5 inch square):
Pine..per cubic foot.. $0.20 to $0.22
Spruce..do.... .16

Pine (Scotch):
Plank (2-4 inch), 3 classes.................................. per 1,000 feet B. M.. 27.00   38.00
Plank (1½-1? inch), 3 classes..do.... 26.00   31.00
Flooring (1 inch), 3 classes...do.... 17.00   22.00
Flooring (1½-inch), 3 classes..do.... 20.00   26.00
Spruce, rough boards, not edged (1-5 inch) ....do.... 12.00
Spruce (1½-inch), edged, 12-18 feet....................................do.... 20.00   22.00

Delivered at Berlin.

Oak (clear), 82 cents per cubic foot, or $68 per 1,000 feet B. M.
Elm, 78 cents per cubic foot.
Railroad ties—pine, 15 cents; oak, 30-35 cents.

It will be seen that prices for some grades are as high as and higher than in New York. The manager is expected to secure at least the government rate, and has discretion in conducting the sales to the best advantage of the government. Under certain circumstances sales by contract without auctioneering, and, lately, selling on the stump, are permitted.

The transportation from the woods, as stated before, is usually left to the buyer; rarely does the administration float the timber or cord wood out, or carry it to a depot or wood yard to be sold from there, or engage in milling or other operations. On the other hand, it has been recognized during the last twenty-five years that good roads and other ready means of transportation increase the price of the wood disproportionately. A good road system is, therefore, considered the most necessary equipment of the administration, and an extension of permanent and movable logging railroads is one of the directions of modern improvement. The interesting, important, and practical features to us in the logging railroads are their movable character, being divided into sets of pairs of short (2 to 5 yard) rails (12 to 16 pounds per yard) attached to from two to four cross-ties, wood or metal, the light sets weighing 75 to 100 pounds (heavy sets up to 166 pounds), so that one workman can readily carry them; the ready connection of sets, one hooking at once into the other without separate mechanism, forming a sufficiently satisfactory joint; the simple "climbing switch," which is applied on top of the track, permitting ready transfer from side track to main track and ready relocation. These roads can be readily laid down without much or any substructure and readily relocated. The cost is shown in the following statement:

For a fully equipped road, 21 to 28 inches width, 6 miles length, for rails and ties.......... $9,000
For earthwork, if any, and laying .... 50 to 500
For rolling stock and apparatus 2,500
                                                                                              _____
                                                                                              12,000

Or $2,000 per mile at the highest.

Upon a basis of 800,000 cubic feet (about 7,000,000 feet B. M.) to be transported, it is calculated that the cost of transportation by railroad, stone road, and dirt road will be about as 1 : 2 : 6, the cost on the first being about 3 cents per 1,000 feet B. M. per mile as against 18 cents on dirt roads.

Comparing the cost of construction it is stated that the ratio between corduroy, gravel road (13 feet wide), macadam, and movable track is as 1 : 1.25 : 2.35 : 1.17, placing the last among the cheapest.

A most instructive exhibit at the World's Fair, in many ways, especially at the present time, since the movement for better roads in this country has begun, was the model of the city forest of Goslar, a small town (13,300 inhabitants) in the Harz Mountains, whose citizens, from this piece of property, a spruce forest of 7,368 acres extent, derive not only their pure drinking water, healthful enjoyment in hunting, and refreshing coolness in summer, but also a net income, amounting in round numbers to $25,000 ($3.40 per acre), toward payment of city taxes. This is the result of careful management, which permits an annual cut of 350,000 cubic feet of wood. Of this only 50,000 cubic feet goes into firewood, and 46 per cent, or 160,000 cubic feet, is saw timber, which sells at 10 to 16 cents per cubic foot; while smaller dimensions, poles, etc., sell all the way down to below 4 cents, and firewood at $1.60 for brush to $4.30 for split or round wood per cord. Until 1875 the district was without proper roads. By an effort of the competent manager the city fathers were persuaded to locate and build a rational system of roads on which altogether, until 1891, there was spent for building and maintenance about $25,000. The greatest interest attaches to the statistics carefully gathered by the district manager, Mr. Reuss, since it is always difficult to determine the money value of such an expenditure in dollars and cents.

The proper location of the roads is the most important feature. The roads are ranked according to their importance; the width and manner of finish depend on their rank. Main roads are macadamized; roads of third rank, which are used for occasional hauling of wood, are dirt roads.

These statistics were exhibited in a neat table, as follows:

STATISTICS OF ROAD SYSTEM IN FOREST DISTRICT OF CITY OF GOSLAR (HARZ MOUNTAINS, GERMANY).

Properly located, graded, and built roads reduce cost of logging and hauling, and advance the price for wood. Area, 7,368 acres spruce forest; annual cut, 350,000 cubic feet; road building begun in 1875; total mileage of improved roads in 1891, 141 miles; cost of road system and maintenance until 1891, $25,000.

Cost of logging reduced by good logging roads.

[Daily wages remaining constant at 60 cents.]

Year.	Length of well built logging roads.	Cost of logging per 100 cubic feet.
	Miles.	
1877	7.5	$1.93
1878	12	1.61
1879	27	1.54
1880	37	1.45
1881	46	1.49
1882	50	1.23
1883	52	1.15
1884	54	1.23

Saving per 100 cubic feet $0.70
Saving on annual cost of 350,000 cubic feet 2,450.00

Cost of haulage reduced by good wagon roads.

[Price per load remaining constant at $3.60 Full load, before improvement, 85 100 cubic feet; after improvement, 175-250 cubic feet.]

Year.	Cost of haulage per 100 cubic feet.
1871-1877, before road improvements	$1.52
1878-1884	.98
1885-1891	.80

Saving per 100 cubic feet .... $0,72.0)
Saving on annual cut of 350,000 cubic feet 2,520.00

Price of wood influenced by road improvements.

[Comparison of prices paid at Goslar and at other Harz districts.]

Year.	Length of improved wagon roads.	Prices for wood per 100 cubic feet.		
		At Goslar.	At other Harz districts.	Difference in favor of Goslar.
	Miles.			
1877	22	$8.25	$8.18	$0.07
1878	31	8.65	8.04	.61
1879	42	9.59	8.44	1.15
1880	55	9 79	8.14	1.35
1881	64	9. 05	7. 78	1.27
1882	68	8. 45	7. 43	1.02
1883	71	8 65	7. 63	1.02
1884	77	10. 17	8. 18	1.99
1885	78	9. 88	8 24	.64
1886	79	9. 59	9 39	.20
1887	81	11. 12	9. 71	1.41
1888	82	11. 12	9 98	1.14
1889	83	11. 39	10. 58	.81
1890	85	11. 72	10 92	.82
1891	87	13. 13	11. 80	1.33
Average for fifteen years		9. 91	8 98	.93

Increase in price on total cut of 350,000 cubic feet $3,255
Total profit from improved road system in reduced cost of logging and hauling, and in advance of price received for wood, per annum 8,225
Or nearly 33 per cent on investment.

Saving their cost in two years.

Cost of road, macadamized in 1885, $6,960; maintenance for one year, $480; total, $7,410. During 1885-86 hauling 470,000 cubic feet requiring on old road 4,273 loads of 110 cubic feet average, at $3.60, $15,282.80 (or $2.70 per 1,000 feet B. M.); on improved road, 2,652 loads of 177 cubic feet average, at $3.60, $9,547.20 (or $1.70 per 1,000 f..t B. M.), saving of $1 for every 1,000 feet B. M. Total saving in haulage, $5,735.60, or 77 per cent on cost of road in one year.

YIELD PER ACRE.

The amount of timber cut per acre is very large as compared with average yields in wild woods. Of late the average yield has varied from about 5,500 cubic feet per acre in Prussia to 9,000 cubic feet for the Saxon State forests. The yield has been steadily increasing since the beginning of this century, and in most States it has been nearly doubled through better management. At that earlier time much land was badly stocked or devoid of any cover, much timber was injured and stunted by continual removal of the litter and consequent impoverishment of the soil, and in most forests the young timber occupied much more than its share of ground, and thus less timber grew. In every one of the States and districts these conditions have been changed materially for the better, the cut was increased from year to year, the wood capital or standing timber grew in total amount, and the productive capacity of the forest soils has generally improved. The cut for any given province or State is generally given as so much per acre of total area. Thus the cut for Saxony is placed at 90 cubic feet per acre of total forest area, though, of course, the yield of those tracts actually cut was about 9,000 cubic feet per acre cut. In the following table the figures relating to the State forests are from recent official records, also those of the corporation forests of Baden, Alsace-Lorraine, Bavaria, and parts of Wurttemberg, while the figures for private forests and most of the corporation forests are estimates based on the experience of former years and of only part of the provinces.

Yearly cut per acre in the State and other forests of Germany (in million cubic feet.)

State.	Cut per acre of forested area		
	Total (including stump and branch wood where used).	Woodover 3 inches (no stump wood).	Timber and bott size material (not fire wood).
For the entire Empire	55		16.4
State forests of—			
Prussia	54	42	19
Bavaria	72	55	24
Wurttemberg	81	67	36
Saxony	90	68	54
Baden	74	59	24
Alsace-Lorraine	57	46	22
Hesse	75	52	16
Mecklenburg-Schwerin	61	50	11.6
The entire Empire	63	43	22.5
Corporation forests of the entire Empire *a*	56	44	16.6
Private forests of the entire Empire *b*	50	30	12

a Partly from official records, part estimate.
b Generally estimated, as no accurate data are available for any entire State.

Using the above basis, the total annual cut of the country (in million cubic feet) is about as follows:

State.	Total cut.	In the forests belonging to—		
		States.	Corporations.	Individuals.
Entire Empire	1,910	710	370	830
Prussia	1,054	331	178	545
Bavaria	354.5	153	44.5	157
Wurttemberg	69.5	38	25	26.5
Saxony	67.3	37.5	3.3	26.5
Baden	85.9	16.6	47	22.3
Alsace-Lorraine	85.3	21.3	23	11
Hesse	34.8	12.7	12.1	10
Mecklenburg-Schwerin	30.7	15	4.7	11

CONSUMPTION OF WOOD MATERIALS.

Thus Germany has a steady and increasing supply of over 1,900 million cubic feet of timber per year (about one-tenth of our consumption) from the lands which in most other countries remain barren wastes. Of these 1,900,000,000 there are near 600,000,000 cubic feet of saw timber and the like, the rest being cord wood and mostly firewood. From this it would appear that Germany produces about 40 cubic feet of wood per head of population, and that of this about 12 cubic feet are saw timber, etc., as against 350 and 50 cubic feet for our consumption. But in spite of the great economy of wood this amount of home-raised material does not satisfy the demand of the home markets, and Germany with its 1,900,000,000 cubic feet is to-day the second greatest importer of wood, particularly of saw timber, in the world.

The import in this case means the excess of import over export, since naturally in all countries an export of some timber takes place.

Consumption of wood (million cubic feet).

Country.	Total.	Produced at home.	Imported	Log timber, etc.		Per cent imported.	Relative importance as import ers.
				Produced at home.	Imported.		
Germany	2,090	1,910	180	570	180	24	40
England	591	140	451	42	451	99	100
France	1,175	1,075	100	300	100	33	22

Per head of population, and comparing with the consumption in the United States, this becomes:

Consumption of wood per capita of population (cubic feet).

Country.	Total.	Produced at home.	Import over export.	Log timber.	Relative wood consumption per head.
					Per cent.
Germany	44	40.5	3.8	15	12.7
England	15	3.6	11.5	13	4.3
France	32	30	2	8.3	9
United States	350	349.7	0.3	a 50	100

a This refers to lumber or sawed material alone.

Since the consumption by sawmills of large timber, particularly coniferous material, is still increasing, it is clear that Germany has not nearly as much forest land as it needs, or else must still improve greatly its methods of production. At present 26 per cent of its saw timber, etc., is imported.

The following figures give an idea of the extent and distribution of the German trade in woods and wood manufactures:

Germany's trade in wood and wood manufactures, 1892.

Country.	Imports.	Exports.
United States	a $2,418,000	$1,504,000
Russia	b 26,808,000	741,000
Austria-Hungary	c 16,363,000	1,946,000
Sweden	d 5,222,000	305,000
France	1,796,000	3,405,000
England	1,314,000	13,448,000
Holland	822,000	2,546,000
Norway	849,000	176,000
Belgium	730,000	1,469,000
Denmark	56,000	967,000
Hamburg	124,000	1,551,000
Switzerland	224,000	1,822,000
East India	e 1,114,000	174,000
Spain	f 1,302,000	354,000
Argentina	g 259,000	129,000
Brazil	68,000	384,000
Porto Rico and Cuba	h 352,000	
Total	60,016,000	30,922,000

a Lumber.	c Oak, etc., logs.	e Largely rattan.	g Largely quebracho.
b Pine logs.	d Sawed lumber.	f Nearly all cork.	h Mahogany, etc.

The prices paid by Germany have so far been very reasonable. Thus her imported lumber cost in 1892 only $18.30 per thousand feet; firewood only $6.50 per cord; fine hewn timber (mostly hard pine in long pieces) $30 per thousand feet, etc.

With the enormous resources in European Russia and Sweden, part of which are not even organized as yet, there is no apprehension of rapid advances in prices and no likelihood of scarcity of supply.

FINANCIAL RESULTS OF FOREST MANAGEMENT.

Concerning the financial results of forest management only the records of the State forests are accessible. It is clear that the income depends on the amount of timber cut and the prices obtained. If, therefore, the yearly cut has been increased, in some cases doubled, by good management since the beginning of this century, the income naturally is doubled. To this increase in amount of salable material there was added a general advance in prices, partly due to the depreciation of money in general, but vastly increased by the improvements in transportation, for which large sums have been expended, especially during the last fifty years.

The financial results of the various Government forest administrations vary considerably, as is natural, since market conditions vary much. It is believed that all these administrations are less profitable than they might be, being managed with great conservatism, and less for greatest financial result than for desirable economic results.

The following table exhibits in a brief manner the results of this kind of management, the figures referring to conditions in 1890 or thereabout. The record for the city of Zurich is added to show how an intensively managed small forest property under favorable conditions of market compares with the more extensively managed larger forest areas:

Forestry statistics of certain German forest administrations, showing average cost of administration, gross and net income per acre, 1890.

States.	Forest area.	Total expenditure.	Revenue.		Expenditures and revenues per acre of forest.						
					Expenditures.						
			Gross	Net	Total.	Per cent of gross income.	Administration and protection.	Marketing crop.	Cultivation.	Roads.	Net revenue.
	Acres.										
Prussia	6,000,000	$8,000,000	$11,000,000	$6,000,000	$1.33	58	$0.48	$0.30	$0.14	$0.06	$0.96
Bavaria	2,300,000	3,150,000	5,880,000	2,730,000	1.37	53	.64	.37	.11	.11	1.19
Wurtemberg	470,000	1,025,000	2,260,000	1,235,000	2.17	45	.87	.92	.22	.33	2.63
Saxony	416,100	1,040,000	2,750,000	1,710,500	2.50	37	.63	.81	.11	.21	4.11
Baden	235,000	404,000	1,070,000	666,000	1.54	40	.32	.83	.15	.12	2.99
City of Zurich	2,760	14,000	26,000	12,000	5.00	54	1.14	2.10	.16	1.14	4.40

The latest figures (1897) show a considerable increase in all directions, expenditures, gross, and net income, over those prevailing ten years ago, and, as we will see further on in the discussion of the conditions in the single States, these increases have been steady for a long period.

The following figures represent the income and expense for State forests of the entire Empire and for the principal States as at present:

Financial results, 1897.

[Million dollars.]

State forests.	Gross income.	Total expenses.	Net revenue.
Germany a	39,361	18,833	20,528
Prussia	17,445	9,079	8,366
Bavaria	8,100	3,881	4,219
Wurtemberg	3,019	1,224	1,795
Saxony	2,865	1,032	1,833
Baden	1,537	618	919
Alsace-Lorraine	1,522	752	770
Hesse	810	405	435
Mecklenburg-Schwerin	609	356	253

a This item is a trifle below the truth, as the small principalities are here assumed to have no larger income than the average of the larger States.

From this statement it appears that Germany has a yearly gross income of nearly $40,000,000 from its State forests, i. e., from one-third of its total forest area alone, while the value of its forest products from the entire forest area (35,000,000 acres) may be estimated to sum up the handsome total of over $107,000,000, or round $3 gross income for every acre under forest cover.

The following table illustrates the results of forest management in the several States. For comparison the figures represent the yearly income and outlay per acre of total forest area, so that for instance the gross income of $3.47 per acre for Germany means that the German State forests yield each year about that sum for every acre of State forest, or $39,300,000 on the whole.

Yearly income and expenses per acre of forested area.

State forests.	Cut of wood per acre.	Gross income.	Expenses		Net revenue.
			Total.	As a per cent of gross income.	
	Cubic feet.				
Germany a	82	$3.47	$1.66	48	$1.81
Prussia	54	2.66	1.38	52	1.28
Bavaria	72	3.71	1.78	48	1.93
Wurtemberg	81	6.50	2.64	40.5	3.86
Saxony	90	6.90	2.36	34	4.54
Baden	73	5.82	2.69	46.2	3.13
Alsace-Lorraine	57	4.21	2.19	49.4	2.12
Hesse	75	4.95	2.37	48	2.58
Mecklenburg-Schwerin	61	2.52	1.47	58	1.05

a This figure represents the average for 90 per cent of all State forests, and would be little changed if data for the other 10 per cent were accessible.

From these figures it appears that the several governments expend on an average about $1.66 per acre per year on their forest property, and that they obtain thereby a gross income of $3.17 per acre and a net revenue of $1.81, or 52 per cent of the gross income per acre per year. Considering the $1.81 as the interest on the value of the forest lands, and using the 3 per cent interest rate as customary for large investments, these figures show that by proper management the German States keep their poorest lands at a capital value of over $60 per acre; in other words, that the German State forests pay $19,000,000 for labor and taxes, and in addition pay interest at 3 per cent on a capital of $60 per acre. A large part of this land if deforested would not support a farmer and would rapidly degenerate into mountain pasture and heath, which at best could not be sold at over $5 per acre, and even then would prove more a detriment than advantage to the community. It also appears from the above figures that the revenue is largely in proportion to the expenses, that the forest which is best cared for also pays the best. The same conclusion is reached by a study of the past. In 1850, when the total expenses per acre in the Prussian forests were only 37 cents, the net income was only 46 cents; to-day it is $1.38 and the net income $1.28, and the same holds for other States. Thus Saxony expended 80 cents an acre per year in the beginning of this century and received 95 cents net income; to-day she spends $2.36 and receives $1.54, or nearly fivefold. That these advances are not merely the expression of higher prices for wood is clear from the fact that the average price of wood for the Prussian cut (300,000,000 cubic feet) has advanced since 1850 from $3.27 per 100 cubic feet to only $4.40, or 37 per cent, while the net income rose from 46 cents to $1.28, or 176 per cent.

Since so much has been argued as to the impossibilities and impracticability of employing these better forestry methods elsewhere, and especially since the idea of sowing or planting forests has at all times been ridiculed in the United States, it may be of interest to note just how Germany expends her money in the woods.

The following figures present the various large items as per cent of the gross income. Thus the total expenses in the Prussian forest use up 50 per cent of the gross income, the logging alone 11.8 per cent, etc.

The expenses represented the following proportions of the total income in per cent:

State forest of—	Total expenses.	Administration and protection (mostly salaries).	Cutting and moving the timber.	Planting, sowing, drainage work, wood roads, etc.
	Per cent.	Per cent.	Per cent.	Per cent.
Prussia	52	21	11.8	7.5
Bavaria	48	21	20	6.6
Wurttemberg	40.5	12	14.6	8.6
Saxony	31	12	14.5	6.4
Baden	46.2	9.4	17.7	10.4
Alsace Lorraine	49.4	17	15.2	8.4
Hesse	48	19	21	3.7
Mecklenburg Schwerin	47	17	17.5	9.2

The above figures are doubly interesting, since they show that in Saxony, the very State where the timber is usually cut clean and the land restocked entirely by planting it with nursery stock, the item of planting, etc., uses up the smallest per cent of the total income—6.4 per cent.

From this brief outline it will be apparent that forestry in its modern sense is not a new, untried experiment in Germany; that the accurate official records of several States for the last one hundred years prove conclusively that wherever a systematic, continuous effort has been made, as in the case of all State forests, whether of large or small territories, the enterprise was successful; that it proved of great advantage to the country, furnished a handsome revenue where otherwise no returns could be expected, led to the establishment of permanent woodworking industries, and thus gave opportunity for labor and capital to be active, not spasmodically, not speculative, but continuous and with assurance of success. This rule has, fortunately, not a single exception. To be sure, isolated tracts away from railroad or water, sand dunes, and rocky promontories exist in every State, and the management of these poor forest areas costs all the tract can bring and often more; but the wood is needed, the dune or waste is a nuisance, and the State has found it profitable to convert it into forest, even though the direct revenue falls short of the expense.

FOREST ADMINISTRATION.

The care and active legislative consideration of the forest wealth dates back fully three centuries. The so called "Forstordnungen" (forest ordinances) of the sixteenth and seventeenth centuries laid the foundation for the present system, and in some States, like Wurttemberg, were never repealed, but merely modified to adapt them to modern views of political economy. The end of the seventeenth century brought much discussion into the subject of forest legislation, as in all other public affairs, and even conservative Germany was led beyond the point of equilibrium, and in most States the State supervision, especially of private forests, was abandoned. This led to the division and parceling of forest properties, and with the diminutive holding came mismanagement and to considerable extent the complete devastation. This condition never affected any of the State forests nor the majority of corporation forests, so that these properties continued on their way to improvement. The wretched condition of many of the private forests is deplored, exposed, discussed, but so far those States which gave the private forest free have been unable to do more than to teach by example and to encourage, both means entirely ineffective when, as is usually the case, the owner is too poor to handle a forest. What remains to be done is being done as fast as means and opportunity offer. The State buys these half wastes, restocks them at great expense, and thus public money pays for public folly.

To provide for a suitable and efficient forest service Germany has expended large sums in promoting forestry education. At nine separate colleges men are prepared for this work, and the forest manager ("Oberfoerster," "Revierfoerster") in any of the State forests is a college-bred man with a general education about equivalent and similar to that leading to a degree of bachelor of science in our better universities. The organization in all German States is similar—a central office at the seat of government, manned by experienced foresters, acts as advisor to the government, shapes the forest policy of the State, introduces all large measures of reform, etc., and acts as court of appeal in important forest cases. In each province, if the State is large (if not, the central office acts), a provincial forest office sees after the work of the province. This office cooperates with the forest managers in preparing plans for every piece of forest land, in determining the cut of the year, and it also examines the work as well as the records of every district, and acts as tribunal for the province in forest matters. But the real managers of the forests are the "Oberfoerster" or "Revierfoerster," each of whom has on an average about 10,000 acres of forest land for which he acts as responsible director. He lives in the forest, keeps himself informed as to all details, plans for every piece of ground (his plans must be approved by his superiors), and executes all plans. He determines where and when to cut, to plant, to build roads, and it is he who sells the forest products. In all cases he has a number of assistants and guards who act as police, and at the same time as foremen to the laborers, directing their work and keeping their time, or measuring their cut or work. The district which the Oberfoerster manages forms the unit in all records and transactions. All forest officials of any responsibility are employed for life or good behavior, their requirements, duties and rights, rates of pay, pension, etc., are all clearly set forth in the forest laws of every State.

In the following pages the conditions and results of forest management in the leading States are fully set forth, based upon the latest official data available.

FOREST MANAGEMENT OF LEADING STATES.

PRUSSIA.

The Kingdom of Prussia, with its 30,000,000 people and an area of nearly 90,000,000 acres of land, representing all natural conditions from the low coast plain to the precipitous mountain system, with its busy centers of manufacture and commerce and its distant rural provinces, stands out to-day as the strongest example of the great benefits of scientific forestry.

The forests of Prussia cover 8,192,505 hectares (about 20,300,000 acres), or 23.5 per cent of the total area. This proportion of forest varies for different parts of the Kingdom from 16 per cent to 39 per cent; it is below the average of 23 per cent in seven provinces, of which only Schleswig-Holstein falls below 16 per cent, and is above the average in six provinces, some of which, like Brandenburg, belong to the densely populated portions of the Kingdom. The area relations have remained practically constant for about thirty years, there being then as now in forest 20,000,000 acres; cultivated 42,000,000 acres, or about twice as much cultivated land as forest.

Of the forest area, 8 per cent belongs to the crown, 30 to the state, 12.5 to villages or municipalities, 1 to Stiftungen (Fonds), 2.7 to corporations, and 52.9 to private owners. This ownership relation has changed a trifle during the last twenty years, the state and municipal forests having gained a little over 1 per cent at the expense of the private and corporation forests.

Situated between latitude 49° to 55° N. and longitude 23° to 40° E. and occupying portions of the extensive coast plain along Baltic and North seas, as well as covering parts of nine separate mountain chains, the forests of Prussia naturally display considerable variety. Of the total 20,000,000 acres, about half falls to the plain, one-fourth to the hilly, and one-fourth to the regular mountain districts. The climate is moderately cold; the mean or average temperature for summer is about 60 to 65° F., varying but little for the different parts of the Kingdom, and being quite uniform for all three summer months. Spring and fall, the latter a trifle warmer and more even than the former, have a mean temperature of about 45° F., while that of the winter months is generally near the freezing point, the coldest weather for any one place and month being rarely below 25° F.

Prussia is a moderately humid country. The records from thirty to seventy years indicate an even distribution of precipitation, varying generally between 22 and 28 inches, reaching a height of over 32 inches, and only 3 out of about 10 stations. With regard to the manner of management, the kind of timber raised, and the financial results of the work, the State forests, for which alone exact statistics exist, may serve as examples, though the results are somewhat better in these than in the forests of municipalities and private owners.

The total area of State forest in 1893 was 2,464,757 hectares, or about 6,750,000 acres. This total area has remained almost unchanged for over thirty years. During this time many large and small tracts have been sold or exchanged to round off the State holdings and to satisfy private rights, many of which had become extremely troublesome and proven a great hindrance in the proper management of the woods. These sales and exchanges were fully balanced by purchases, especially of poor, unproductive private forests and heath lands, for which purpose of late the State appropriates annually the large sum of 1,000,000 marks ($250,000), the policy of increasing the State holdings having been steadily pursued for more than fifty years. About two-thirds of the State forests are situated in the North German plain, though some occur in every province of the Kingdom.

Of these State forests 97 per cent are regular timber forest, mostly pine and spruce, where the final crop is intended to furnish saw timber, and every particular parcel is supposed to be stocked with trees of nearly the same age. Only one-half of 1 per cent is managed as "Plenter-wald" with the method of selection where trees of all sizes and age mingle together on the same parcel and the logging merely involves the selection of suitable sizes. One-half of 1 per cent is standard coppice, where the bulk of the trees, commonly hard woods, are cut off while still small, 15 to 30 years old, while a small portion is left over to grow into larger sizes; and 1.7 per cent is managed as coppice, largely oak coppice for tanbark, where the trees (only the sprouting hard woods) are cut down every ten to twenty-five years, the wood being utilized chiefly as poles and fuel. Of the timber forests, 62 per cent is stocked with pine, almost entirely Scotch pine (*Pinus sylvestris*), furnishing hard pine similar to our red or Norway pine, 16 per cent is beech, 12 per cent spruce, and nearly 6 per cent oak forest. Thus about 75 per cent of all Prussian State forests are coniferous woods and only about 25 per cent stocked with hard woods, principally oak and beech.

In general the trees of the timber forests are cut at an age of about 100 years (a 100-year rotation). At present 13 per cent of the area is stocked with trees over 100 years old; 13 per cent, 81 to 100 years old; 14 per cent, 61 to 80 years old; 18 per cent, 41 to 60 years old; 19 per cent, 21 to 40 years old; 19 per cent, 1 to 20 years old, and about 4 per cent are cut clean (recent fellings) to be reforested at once.

SAXONY.

If Prussia may be regarded the best example of the success of rational forestry in a large country, and Wurttemberg can be cited as proving the great value of a very conservative, almost paternal, attitude of the State with regard to its forests, surely Saxony deserves the credit of leading all other countries in the intensity of its forest management.

The total area of the State is 3,700,000 acres, and its population 3,182,000, and its total forest area about 1,020,000 acres, or 27 per cent. Of this forest area, 173,889 hectares, or nearly 430,000 acres, equal to about 43 per cent of all forests of the country, belong to the State. The accurate records for these State forests have been kept for more than eighty years, and fully illustrate the development and growth of forestry in the Kingdom. The bulk of the forests are mountain forest; 91 per cent in conifers, mostly spruce, and only 9 per cent in hard woods, most of which is beech; while only about 4 per cent is nonproductive rock and water area.

As early as 1764 the State of Saxony began the improvement of the then rather dilapidated forest properties. The real systematic work of forest survey and management, however, did not begin until Heinrich Cotta (often called the father of modern forestry) began his noteworthy efforts in 1811. Though the Government never appropriated special funds for the increase of its forest holdings, the money which accrued from the sales of other State lands, as well as roadways, building sites, etc., sufficed to increase the area during the past eighty years by fully 16 per cent, the growth being a slow, steady one, fully illustrating the policy of the Government.

Thus the growth was: 1836 to 1846, 5,000 acres; 1846 to 1853, 5,000 acres; 1853 to 1863, 5,000 acres; 1863 to 1873, 17,200 acres; 1873 to 1883, 17,200 acres; 1883 to 1893, 12,500 acres.

As in all German States, nearly every piece of State forest was burdened by rights of private persons and corporations, for which Saxony has paid, almost entirely in cash, the handsome price of $1.300,000.

During the last sixty years the area stocked with conifers has steadily grown from about 310,000 to over 385,000 acres, and the area of beech and other hard woods except oak has been proportionately diminished, the hard woods all told covering at present only about 14,000 acres, or a little over 3 per cent of the forest area. The condition of the forests, though, of course, very good at the start, if compared to ordinary wild woods, has steadily improved since 1817, in spite of the fact that each decade a larger amount of wood was cut.

The following figures serve to illustrate this important fact and at the same time show that there has not only been a steady increase in the total amount of wood standing and the amount cut, but that the larger sizes form to-day a much greater per cent than formerly:

Years.	Total amount of wood cut each year (average for each decade).	Per acre of forested area		
		Amount cut.		Amount standing per acre on total area.
		Total.	Wood over 3 inches thick (cord wood and timber).	Timber (not cord wood).	
	M. cub. ft.	Cubic feet.	Cubic feet.	Cubic feet.	Cubic feet.
1817-1826	21,400	60	40	7	
1827-1836	21,800	61	39	10	
1837-1846	20,400	56	36	11	
1847-1853	23,500	64	44	14	2,120
1854-1863	26,000	70	48	23	2,280
1864-1873	31,600	82	60	37	2,180
1874-1883	36,600	90	66	47	2,650
1884-1893	37,400	90	68	54	2,620

From these figures it appears that the cut on the whole has increased from 21.000,000 cubic feet to 37,000,000, or by fully 57 per cent, and the cut per acre and year of total forest area from 60 cubic feet to 90 cubic feet, or exactly 50 per cent. Moreover, of the 90 cubic feet per acre in 1893 there were 68 cubic feet, or 75 per cent, wood over 3 inches (excluding stump wood), while from 1817 to 1826 only 66 per cent was over 3-inch stuff. But what indicates even more strongly the effect of better management is the fact that more than half of the cut of 1893 was sold, not as cord wood, but as timber (saw timber, etc.), while even as late as 1865 only a fourth could thus be utilized, though the manner of selection (inspection) has changed but little since that time. That with all this intense utilization of the forest the standing timber should increase instead of becoming exhausted is perhaps the strongest example of the success of scientific forestry and one which in this country would scarcely be believed possible by most of the lumbermen and woodsmen.

Practically, all State forests are timber forests and the prevalent method of treatment has for a long time been the "kahlschlag" method of cutting, where all trees are cut at the harvest and the bare area is at once planted with nursery stock. The expenses for cultural work all told, including maintenance of nurseries, seed and plant purchases, as well as planting, amount to only 12 cents an acre per year, or 1.8 per cent of the gross income, while for the last twenty years more than twice this sum has been expended for construction and improvement of roads, the great value of which are nowhere more fully recognized than in busy Saxony.

The financial results are exhibited in the following table:

General financial results in the State forests of Saxony.

Years.	Annual income (gross).	Annual expense.	Annual net income.	Per acre and year of total forest area.		
				Income (gross).	Expense.	Net income.
1817 1826	$649,000	$297,000	$352,000	$1 75	$0.80	$0.95
1827 1836	682,000	321,000	371,000	1.80	.86	1.00
1837 1846	761,000	342,000	419,000	2 02	.90	1.12
1847 1853	976,000	388,000	588,000	2.56	1.02	1.54
1854 1863	1,368,000	443,000	925,000	3.53	1.14	2.39
1864 1873	1,986,000	561,000	1,425,000	4 91	1.39	3.52
1874 1883	2,624,000	875,000	1,749,000	6.23	2.08	4.15
1884 1893	2,830,000	996,000	1,834,000	6.66	2.29	4.37

The extraordinary results indicated in the above table can not entirely be credited to the increase of wood prices and the general depreciation of money during this century; they are primarily the monetary expression of the improvements indicated in the previous tables; they mean increased sales, and sales of older, larger, and better material.

When it is considered that Saxony has taken in about $190,000,000 during the last fifty years from a small area of rough lands (left waste in many countries, even in Europe), a tract of land half the size of a good county in Wisconsin, the great advantage of a careful treatment of forest areas must become clear to everyone. Considering the net income as the interest of the value of the forest lands at the prevailing 3 per cent rate, the table shows that scientific care has increased the value of these poor mountain lands from $100 to $150, whereas their deforestation would quickly convert them into poor alpine pastures which would bankrupt their owners at $10 an acre. The table also shows clearly that it is not accident, not merely a general improvement of the country, but that it is careful, systematic work which has led to these improvements. When Saxony spent only $1 on each acre of forest land she received only $1.54 net income; when she spent $2.39, her net income was more than doubled, reaching during the ten years ending 1893 $4.37.

The following figures illustrate the nature and relative importance of the expenses per acre as compared with the income, as well as the prices obtained for the material:

Decade ending—	Price per cubic foot of wood over 3 inches.	Wood cut.	Gross income.	Total.	For administration and protection	Felling and moving timber, etc.	Planting and other cultural work.	Roads.
	Cents.	Cubic feet.			Cents.	Cents.	Cents.	Cents.
1826	4.2	60	$1 75	$0.80	28	30	8	2
1836	4.7	61	1.86	.86	40	31	8	5
1846	5.6	56	2.02	.90	44	31	10	4
1853	6.0	61	2.56	1.02	47	37	11	5
1863	7.1	70	3.53	1.14	49	45	13	6
1873	8.1	82	4.91	1.39	54	62	10	11
1883	9.4	90	6.23	2.08	77	92	13	24
1893	9.9	90	6.66	2.29	93	95	14	26

From the above it appears that the prices of wood have doubled since 1817, but that during the last twenty-five years they have remained practically constant. Part of this advance is due to the general advance of prices, but part also to the improvement of the material sold. The advance in the expenditure for administration since 1846 is due both to the advance in wages and salaries generally (seen also in the advance of cutting expenses), but is also due to the greater competence of the administration. Saxony, unlike Michigan and other States of this Union, prefers to spend the money in protecting its forest rather than saving the expense and losing the property. Of special interest is also the fact that even in this intensive management, where almost every acre is reforested by planting with nursery stock, the cultural operations, including drainage and kindred expenses have varied only within a few cents per acre, involving during the last thirty years generally less than 2 per cent of the gross income. To many in this land of forest fires it may perhaps be remarkable that this general enemy and its destructions have not been of sufficient consequence to deserve compilation for this general statement. These mountain forests of spruce and pine are simply not allowed to burn up.

The management of the forests of Saxony is similar to those of Prussia. While those of the State are under conservative and most efficient care, those of private persons and corporations are practically free; the only thing the State authorities do is to give good example, assist private individuals, etc., by furnishing cheap plant material from the forest nurseries and to prepare plans for the management of forests if such plans are asked and paid for.

BAVARIA.

The kingdom of Bavaria has a total area of about 18.8 million acres, or little more than half that of the State of Wisconsin, supporting a population of about 5,589,000 people. It comprised about 10,500,000 acres, or 56 per cent, of fields and gardens; 750,000 acres, or 4 per cent, of pasture lands; 6,350,600 acres, or 34 per cent, of forest; 1,200,000 acres, or 6 per cent, of unproductive land, largely mountains, roads, and water surfaces.

On the whole, this relation of areas has not changed materially in over thirty-five years, so that in 1893 the total area of forest lands is given at about 6,200,000 acres, or at 35.1 per cent of the entire land surface.

Of these 6,200,000 acres there are: State forests, 2,160,000 acres, or 34.8 per cent; corporation forests, 780,000 acres, or 12.6 per cent; pond forests, 110,000 acres, or 1.7 per cent; private forests, 3,150,000 acres, or 50.9 per cent.

The forest laws and forest organization resemble those of Baden and Wurttemberg. The private forests are under State supervision, clearing of forest lands requires a permit, the mismanagement or devastation of a forest property is forbidden, and devastated forest areas are to be reforested by the State and the expense charged to the forest. All corporation and Fonds forests are under direct control of or are managed under control of the State forest authorities, so that fully one-half the forest area of Bavaria is under careful treatment. As with all German States, Bavaria constantly endeavors to increase the State holdings, and deteriorated and other forest properties are bought up as opportunity offers. During the fifty years ending 1894, the State purchased about 144,000 acres, at a cost of $5,577,000, or about $38 per acre. Besides this increase of territory, the State has, during this same period, expended about $3,800,000 in the purchase of easements or servitude, involving 10,716 separate cases of privileges to timber and firewood. Nevertheless, there are still many of these privileges or servitudes, which require an annual outlay of over $100,000 and thus represent a capital value of over $10,000,000.

The distribution of the forests over the kingdom is rather an even one. Six of the eight provinces have over 30 per cent, the lowest 22 per cent of forest area, while the highest 38 per cent. Of the entire forests area about 90 per cent is covered by timber forest, where the timber is cut usually at about 100 to 120 years, and only 9.4 per cent as coppice and standard coppice.

Forty years ago the same was stocked as follows:

	Timber forests.	Coppice and standard coppice.	Selection timber forests.
	Per cent.	Per cent.	Per cent.
State forests	92	5	3
Corporation forests	62	35	3
Private forests	72	13	11
True average for the whole area	78.5	12.7	8.4

The principal forest trees are the conifers, chiefly spruce. Of the total, about 46.2 per cent is spruce and fir, 30 per cent pine, 9.7 per cent beech, 4 per cent oak (two-thirds oak-bark coppice), 2.3 per cent other hard-wood timber, 6.8 per cent other hard-wood coppice.

Thus, conifers represent about 77 per cent, the hard woods 23 per cent. The conifers are primarily the trees of the mountains, the hard woods, beech particularly, being most abundant in the valley of the Rhine, the Palatinate, and Lower Franconia, where the beech forests cover as high as 80 per cent of the forest area.

In 1860 the total cut for the kingdom was 275 million cubic feet of stem wood, 35 million cubic feet of branch wood, 30 million cubic feet of stump wood, making a total of 310 million cubic feet, and was divided as follows:

	Per cent of total cut.	Yield per acre.
		Cubic ft.
State forests	39	58
Corporation forests	14	46
Private forests	46.5	47
Total	100	51

For the State forests alone the cut in 1894 of wood over 3 inches, excluding branch and stump wood, was 55 cubic feet per acre, and included saw and other timber, 55 million cubic feet; cord wood (exclusive of branches and stumps), 61 million cubic feet.

The financial results for the 2.16 million acres of State forests were, in 1894: Total income, $8,100,000, or $3.71 per acre; total expense, $3,881,000, or $1.78 per acre; net income, $4,219,000, or $1.93 per acre.

Compared to other small States of Germany, particularly Saxony and Wurtemberg, the net revenue per acre of forest is decidedly low; but it must not be forgotten that a considerable part of these State forests is situated in the high Alps, where the difficulties of removing the timber have so far been very great, and the value of timber consequently very small. Thus, fine timber trees, worth $50 to $100 on the markets of the lower Rhine, are worth little over $1 apiece in these Alpine districts.

As might be expected, the permanent improvements of the forests, particularly the construction of highways and roads, still require large sums every year. Thus, in 1894, Bavaria spent over 1,000,000 marks ($250,000) on road construction.

The management of the forests is quite similar to that of the other German States. The Revierförster, corresponding to the Prussian Oberförster, is the responsible manager of each district. The districts are quite large: they include usually about 5,000 acres of State forest, so that one Revierförster is usually 6 to 10 miles from his neighbor.

For all State and corporation forests, an area of a little over 3 million acres, there are 609 Revierforster or managers, 1,589 guards and assistants, besides 175 accountants and 107 superior officials. The manager or Revierförster makes and executes the plans and keeps the records for the woods of his district.

As in Wurttemberg, rational measures for the proper use and treatment of forests of Bavaria date back to the beginning of the seventeenth century. As early as 1616 a forest law was passed which embodied all that seemed at that time desirable. This law was modified, some complications arising from the change of size and form of the kingdom, and also through the radical views promulgated during the second half of the eighteenth century. On the whole, however, Bavaria remained conservative, which in view of its large mountain forests must be regarded as particularly fortunate.

The establishment of the forest school at Munich took place about 1789, when a general reorganization occurred, and the functions of the forester changed from those of a hunter to those of a producer of timber.

WURTTEMBERG.

This little State, with an area of about 4,820,000 acres, or about one-seventh that of Wisconsin, and a population of little over 2,000,000 people, ranks among the most conservative as well as the most successful among the commonwealths of Europe. In matters of forestry this State began proper measures as early as 1614, when laws were inaugurated for the proper treatment of forest properties, which remain fundamental to this day. These early laws, which made the proper care of forests obligatory to all and forbade both forest devastation and clearing (the latter possible only on permit), were properly enforced and maintained even through the troublesome times of the end of the eighteenth century. They were remodeled and perfected to suit modern conditions in 1875 and 1879 the law of the former date dealing with the forests of public corporations, the latter with State and private forests in general.

The "forest police law" of 1879 requires:

(a) Clearing of forest requires a State permit; illegal clearing is punished with a fine.

(b) A neglected piece of forest shall not become waste land; the State authority sees to its reforestation, with or without help of owner, the expenses to be charged to the forest.

(c) If the forester is convinced that a private owner cuts too much wood or otherwise mismanages his forest, he is to warn the owner, and if this warning is not heeded the forest authority may take in hand and manage the particular tract.

(d) Owners of small tracts of forest can combine into associations and can place their properties with municipal or even State forests for protection and management. In the latter case they share the advantages of part of the municipal or communal forests which are managed by State authorities.

The law of 1875 relating to the management and supervision of forests belonging to villages, towns, and other public corporations places the forests under this category all under direct State supervision; there being a special division of corporation or municipal forests in connection with the State forest bureau. The law demands that all corporation forests be managed in accordance with the principles of a continued supply, the same as the State forests. The corporation may employ its own foresters, but these must be approved by the forest bureau and are responsible for the proper execution of the plans of management. These plans are prepared by the foresters and must be approved by the State forest authorities. If preferred, the corporation may leave the management of its forests entirely to the State authorities. This is always done if a corporation neglects to fill the position of its forester within a certain period after it becomes vacant. Where the State forest authorities manage either corporation or private forest, the forest is charged with 8 cents per acre and year for this administration. This fee is generally less than it costs, so that the State really has been making a sacrifice so far in providing a satisfactory management for those forests.

As in all other German States, nearly every piece of forest land was formerly encumbered with certain rights which entitled the holders to certain fixed amounts of firewood, timber, to pasture live stock, etc. The law of 1848 obliges the holders of these rights to part with them if the proprietor pays the value of the rights, the manner of ascertaining the value being set forth in the law itself. Thus, for the right of cutting his supply of firewood in a forest the holder of the right is paid a sum which if placed at 4 per cent interest will purchase as much wood as the holder of the right used per year, the average of twelve seasons being the criterion. Of the different rights or privileges, those concerning pasturage and the cutting of hay in the forests are practically settled, and the State paid between 1873 and 1880 about 2,445,000 marks, or $611,000, for these rights. For privileges of cutting wood and timber the State has expended large sums. Even prior to 1848, between 1825 and 1850, forest land valued in the aggregate at about $3,000,000, and between 1850 and 1880 over $500,000 more have been paid out to rid the woods of these pestiferous rights, and yet as late as 1873 these rights were worth $32,000 per year, or a capital (at 4 per cent interest) of $800,000.

In matters of taxation all forests are assessed according to the net revenue which they produce. Of the total area of the land, about 42 per cent is plow land, 18 per cent meadows and pastures, 30 per cent forest, 3 per cent gardens and vineyards, and 2 per cent roads. In its distribution over the State the forest forms 27 per cent of the area of the Nekar Kreis, 39 per cent of the area of the Schwarzwald Kreis, 31 per cent of the area of the Jaxt Kreis, and 25 per cent of the area of the Donau Kreis.

Of the total of about 1,470,000 acres of forest, 480,000, or 32 per cent, belong to the State; 470,000, or 32 per cent, to corporations, and 530,000, or 36 per cent, to individuals.

Of the corporation forests, nearly 360,000 acres are managed by State foresters; of the private forests, 200,000 acres are held by the nobility, including the royal family.

Accurate statistics have been prepared so far only for the State forests and of late also for the corporation forest, so that a more detailed description of these classes must serve as illustration for the whole.

The State forests of 480,000 acres occupy parts of all four provinces of the country. About 92 per cent lie between 900 and 2,400 feet altitude; 42 per cent are stocked on level ground, 29 per cent on gentle slopes, and about the same amount on steep declines. Over 40 per cent of these forests are situated on sandy soils, and the rest are largely on the poor limestone soils of the Jura, and only a small part on the drift formation skirting the north side of the Alps.

Of the State forest area there is covered by a pine growth of spruce, 28 per cent; beech, 20 per cent; fir, 9 per cent; pine, 7 per cent; mixed growth of conifers, 14 per cent; conifers and hardwoods, 9 per cent; mixed hardwoods with oak, 7 per cent; mixed hardwoods without oak, 2 per cent. Thus about 60 per cent is coniferous growth and only 30 per cent hardwoods, with about 9 per cent mixed timber.

Fully 97 per cent of the State forests are managed by the timber forest system. The rotation is for timber forest, 100 years for 74 per cent of the area; 80 years for 24 per cent of the area, and 120 years for 2 per cent of the area.

At the present (1894) the areas containing timber over 100 years old cover 11 per cent of the area; 81 to 100 years old cover 15 per cent of the area; 61 to 80 years old, 15 per cent; 41 to 60 years old, 17 per cent; 21 to 40 years old, 19 per cent; 1 to 20 years old, 23 per cent; so that a fairly regular distribution for a 100-year rotation exists.

These timber forests yield about 56[1] cubic feet per acre of timber from the main cut or harvest and 11 cubic feet per acre from thinnings, making in all 67 cubic feet per acre and year for the entire area. The 3 per cent managed in coppice and standard coppice cut only about 14 cubic feet per acre and year.

The total cut for 1894 was, for wood over 3 inches thick: Oak, 1,200,000 cubic feet, or 3.9 per cent; beech and some other hard woods, 7,900,000 cubic feet, or 26 per cent; conifers, 21,500,000 cubic feet, or 70 per cent.

This cut was composed of—

A.—*Timber generally over 6 inches at the top end.*

	Amount.	Per cent.
	Cubic feet.	
Oak	560,000	3.8
Other hard woods	420,000	2.8
Conifers	13,800,000	94
Total	14,780,000	100

B.—*Poles 2-6 inches, 3 feet from butt end.*

	Amount.	Per cent.
	Cubic feet.	
Oak	1,500	0.2
Beech and other hard woods	6,400	.9
Conifers	685,000	99
Total	692,000	100

C.—*Cordwood.*

	For wooden ware.	For firewood.
	Cubic feet.	Cubic feet.
Oak	46,000	590,000
Beech and other hard woods	78,000	7,400,000
Conifers	295,000	6,450,000

The above figures, especially those for the yield in saw and other timber, clearly point out the great advantage of the conifers over the hard woods. The same is also clearly illustrated by the fact that the material sold as firewood forms only 40 per cent in conifers, but 94 per cent in

_____
[1] This means that if the timber is 100 years old, as most of it is, each acre of forest cuts 5,600 cubic feet of wood at time of harvest.

beech and other hard woods, leaving out the oak. Moreover, the yields have been much greater for conifers than beech.

Thus the yield for material over 3 inches thick in the hard woods was only 51 cubic feet per acre and conifers 74 cubic feet per acre, while the average value of the two is about as 5 for beech and other hard woods, leaving out oak, to 8 for coniferous wood, so that the yield in money per acre for the two was more nearly 2.4 times as great for conifers as for hard woods.

The prices obtained for wood, generally delivered at the main roads, was: Timber, oak (white oak), 25 cents per cubic foot; conifers, 11.7 cents per cubic foot. Cord wood, beech, 4.9 cents per cubic foot, or $6.30 per cord; conifers, 3.6 cents per cubic foot, or $4.60 per cord.

The money results were for 1894 as follows:

Gross income	$3,019,000, or 100 per cent
Total expense	1,224,000, or 40 per cent
Net income	1,795,000, or 60 per cent

or per acre of forest area:

Gross income	$5.20
Expenses	2.51
Net income	3.69

this latter forming 59 per cent of the gross revenue.

Among the expenses were conspicuous:

Felling of timber	$397,000
Administration and protection	339,000
Roads, new, and repair	163,000
Taxes	103,000
Planting, sowing, etc	91,000

The following figures illustrate the progress of the last eighty years, and at the same time indicate how steadily this small area of otherwise almost valueless land has been made to furnish an ample supply of timber and a handsome revenue:

Results of forest management in the State forests of Württemberg.

Year.	Forest area.	Wood over 3 inches thick cut each year.	Price per cubic foot.	Per acre and year.[a]	
				Net income.	Cut wood over 3 inches.
	M acres.	M cubic feet.	Cents.		Cubic feet.
1815				$0.30	
1819	472			.42	
1823		15,200		.52	33
1828	160	17,200		.64	37
1834	445	17,700		.85	39
1841	447	25,000		1.78	56
1845	452	25,400		1.93	55
1850	452	23,800		1.11	52
1855	455	26,600	4.3	1.42	58
1860	457	28,400	7.5	3.22	61
1865	460	25,300	9.7	3.54	54
1870	465	26,600	7.5	2.62	57
1875	467	28,800	10.7	4.21	61
1880	471	28,700	8.0	2.66	60
1885	474	29,400	8.1	2.90	61
1890	476	30,200	8.7	3.33	63
1894	480	30,600	9.3	3.69	63

a Refers to entire forest area—swamp, water, surfaces, and all.

Most of the logging is done by the cubic foot or cord, and the prices are about 60 to 65 cents per 100 cubic feet of coniferous and 80 cents per 100 for hard-wood timber, while cord wood is generally worked up for about $1 per cord, including piling at roadway. All cut-over land is at once reforested. During 1894, 275 acres were thus recovered by seeding and about 6,000 acres by planting, the latter being thus generally the rule, especially in the coniferous districts. The total expenses of cultural work were $88,000, or less than 3 per cent of the gross income.

The thinnings of the dense sapling timber involved during the year about 20,000 acres and furnished about 240 cubic feet of wood per acre. Most of this material in the hard-wood district has to be cut into inferior firewood, but the spruce, fir, and pine can usually be sold as poles and pulp stuff, etc.

Though largely stocked on sandy soils and composed of pine and other conifers, there are no forest fires reported for the year. The administration of forests is in the hands of "Revierfoerster," corresponding to the Prussian "Oberfoerster," who prepare the plans and execute them, being assisted by a body of subalterns. The district of a Revierfoerster covers about 10,000 acres of forest, while the range or "hut" of the forest guard is generally about one-tenth of this. These guards also serve as foremen in all cultural and felling operations, but the Revierfoerster is supposed to keep fully informed on all details and preserve accurate record. Besides their duties as State forest officers, it is expected that these men also keep themselves informed as to the condition of private and other forests.

BADEN.

In this intensively cultivated little State, with a total area of only about 3,720,000 acres, supporting a population of 1,656,000, the forests occupy over 37 per cent of the entire land surface. The forest area has increased between 1880 and 1895 by over 50,000 acres, being in the latter year 550,891 hectares, or about 1,360,000 acres. These forests were owned as follows:

Owner.	1895.	1880.
	Acres.	Acres.
State	237,000	232,000
Villages and towns	620,000	610,000
Other corporations	47,000	33,000
Private persons:		
Nobility	147,000	147,000
Others	310,000	285,000

The forest policy of Baden has been conservative and there is no State in Germany where the general conditions of the forests are better. Since all municipal and corporation forests are under direct State control, being managed by the State forest authorities, about 910,000 acres, or over 60 per cent of all forests, enjoy a careful, conservative treatment, which insures to them the largest possible return in wood and money. But even the private forests are under the supervision of the State authorities, and though the private owner may use his forest very much as he pleases he can in no way devastate or seriously injure it. Clearing requires a permit, also a complete clearing cut, which latter is permitted only if the owner guarantees the reforestation of the denuded area within a given time. Bare and neglected spots in forests must be restocked, and failure of private owners to comply with the forest rules and laws leads to temporary management of the forest by the State authorities, such management never to continue less than ten years. Of the State forests there are about 93 per cent timber forest with a rotation of eighty to one hundred and twenty years and only 7 per cent coppice and standard coppice intended to produce tanbark and firewood. Of the corporation forests about 83 per cent are timber forest, so that of all the forests under State management about 85 per cent are timber forest managed on long rotations and furnishing large returns.

Of the State forests, 21 per cent are hardwoods, with little or no conifers; 30 per cent are mixed forests, hardwoods, and conifers in about equal parts; 49 per cent are coniferous forests, the bulk being stocked with spruce and fir, while only about 4 per cent of the total is stocked with pine alone.

Full and accurate statistics existing only for the State forests, and, as far as the annual cut is concerned, for corporation forests, the following figures apply only to about 60 per cent of the forests of the country.

The cut for 1894 was in—

	State forests.	Corporation forests.
	Cubic feet.	Cubic feet.
A. From timber forests:		
Main crop	11,100,000	29,100,000
Thinnings	4,300,000	3,800,000
Stumps	150,000	330,000
B. From coppice and standard coppice:		
Main crop	750,000	7,600,000
Thinnings	20,000	120,000
Stumps		50,000
	16,560,000	47,000,000

This same cut per acre of total forest area is—

Timber forest: Cubic feet.
  State....... 74
  Corporation 71
Coppice and standard coppice:
  State....... 53
  Corporation 65

This enormous yield of nearly 64 million cubic feet of wood Baden has obtained from this small area for many years without in the least decreasing the amount of standing timber or wood capital. In the State forest the cut per acre since 1867 has never been less than 57 cubic feet per year, or since 1885 has never fallen below 71 cubic feet, while twice since 1870 it has been over 85 cubic feet per acre and year.

Of the total of nearly 64 million cubic feet, 19,200,000 cubic feet are timber and other wood not sold as fire or cord wood, and 29,100,000 cubic feet are cord wood over 3 inches.

The forests of Baden are generally well located, and the State has long realized the great importance of good highways, so that the prices for timber are generally good and the income from the woods correspondingly high.

The following prices in the woods were obtained in 1894:

For round timber long lengths and saw logs (per cubic foot):
  Oak....... $0.16 to $0.39
  Beech .... .15
  Ash and maple .... .24
  Birch .... .08
  Alder .... .23
  Other hardwoods .... .16
  Conifers, long stems....... .07 to .13
  Conifers, saw logs....... .11 to .14
  Conifers, railway ties....... .08
For cord wood (per cord):
  Beech....... 6.50 to 8.40
  Oak....... 5.80 to 10.80
  Other hardwoods....... 6.30 to 7.80
  Conifers....... 4.00 to 4.80

The financial results in the State forests were as follows:

For the year 1894—
  Total income $1,337,000
  Total expenses 618,000
  Net income 719,000
Or per acre of forest-stocked area—
  Gross income......... $5.82, or 100 per cent
  Expenses .... 2.69, or 46.2 per cent
  Net income .... 3.13, or 53.8 per cent

How steadily this handsome revenue has been received may be inferred from the fact that during the twenty-eight years ending 1894 the gross income has never been below $4.24 per acre; that for thirteen out of the twenty-eight years it varied between $4.24 and $5; that twelve years it was between $5 and $6, and three years above $6 per acre.

The following figures show this relation for the period 1881 to 1894:

Production and cost per acre of forested area.

Year.	Cut.	Annual income (gross).	Annual expense.	Annual net income.	The expense is of the income—
	Cubic feet.				Per cent.
1881	59	$4.08	$2.13	$1.94	52
1882	62	4.41	2.17	2.24	49
1883	67	4.80	2.24	2.55	47
1884	67	4.87	2.30	2.57	47
1885	71	5.15	2.34	2.80	45
1886	74	5.23	2.47	2.76	47
1887	75	5.33	2.60	2.73	49
1888	73	5.16	2.50	2.65	49
1889	76	5.48	2.59	2.88	47
1890	80	5.85	2.60	3.25	44
1891	74	5.65	2.58	3.06	46
1892	73	5.73	2.65	3.08	46
1893	72	6.07	2.64	3.42	43
1894	73	5.82	2.69	3.13	46

Considering the fact that these forests, in the aggregate only about as large as ten townships, are scattered over considerable area, and thus their protection and management is rendered much more costly than if in more compact form, these results are certainly most remarkable.

Of the expenses, those of special interest are:

Logging (generally)	$221,000
Administration	132,000
Protection	54,000
Roads, new and repair	77,000
Sowing, planting, etc	42,000

As stated before, wherever the forest is cut, reforestation is at once begun. As in other States, part of this is carried on by the process of natural regeneration, where the old trees are never entirely removed until they have been made to seed the ground, but part is also done by artificial sowing and planting. In 1891 about 125 acres were seeded anew; 65 acres were seeded to correct failures of former years; 760 acres were planted for the first time, and about 850 acres of former failures were corrected.

The work of seeding costs $11.95 per acre, the planting $11.43 per acre, which shows that it is not by a penny-wise and pound foolish system of retrenchment that the most extraordinary results of the Baden forest management are attained.

ALSACE AND LORRAINE.

These two small provinces, formerly under French rule, have an area of about 3,600,000 acres and a population of about 1,500,000, and are under the Imperial Government. The existing forest laws of these provinces left in force on their transfer to Germany, so that now, as in former times, the French "code forestier" of 1827 and some subsequent dates decide in all affairs concerning the forests. The laws in the main are like those of Baden: they restrict the right of the private owner to a proper use of the forest and forbid all devastation; any clearing requires a State permit, and with regard to protection against fire, insects, etc., they are subject to the ordinary forest police regulations. As in Baden, the forests of corporations are managed by State authorities, so that a well-planned forestry system applies to all forests except those of private owners, and even these are under rigid supervision and partial control.

The total area covered by forest is 444,466 hectares, or about 1,100,000 acres, forming about 30 per cent of the entire land surface. Of this forest area there belong to the State 340,000 acres, or 31 per cent; villages and towns, 490,000 acres, or 45 per cent; private owners, 220,000 acres, or 20 per cent. Besides these there are about 40,000 acres of land belonging jointly to the State and villages and 6,000 acres belonging to corporations other than municipalities.

Since all forests, except those of private owners, are under the management of the State forest authorities, fully 80 per cent of the forests of these provinces are in most excellent condition. Though the exact proportion has not been ascertained, it may be said that about 60 per cent of the forests are hardwoods, largely beech and oak, and only 40 per cent conifers.

The total cut for 1891 was—

	Cubic feet.
For State forests	21,400,000
For corporation	33,000,000
Total	54,400,000

of which about 17,500,000 cubic feet was nutzholz, or timber not sold as cord wood or firewood. Of the 21,000,000 cubic feet of wood cut in the State forests there were in 1891:

Kind of wood.	Timber (nutzholz).	Cord and other firewood.	Total of wood.	Per cent of total cut.
	Cubic feet.	Cubic feet.	Cubic feet.	
Oak	1,600,000	2,100,000	3,700,000	18
Beech with other little hardwoods	800,000	8,300,000	9,100,000	43
Conifers	5,500,000	2,700,000	8,200,000	39

The average price per cubic foot was:

For timber or work wood—   Cents.
    Oak 17
    Beech 11
    Conifers.. 8.5
For firewood —
    Oak 5.5
    Beech 6.7
    Conifers.. 4.2

On the whole the State received 7.2 cents per cubic foot for all its timber and firewood. Among the improvements made during the year the items of roadmaking and reforestation are most conspicuous.

In the State forests alone about 1,500 acres were seeded, generally at a cost of $2 to $3 per acre, the lowest being 60 cents; while in few cases the cost exceeded $4 per acre. About 3,200 acres were planted, 1,280 acres for the first time, the rest being corrections of former failures. Planting largely with hardwoods cost on an average about $5.50 per acre. Roadmaking is vigorously pursued, as much of the land is quite rough, and well-planned, permanent, macadamized roads have proven to be among the best investments. In some of the districts forest railways have also been constructed.

The final results during 1891 were as follows:

Income from wood.. $1,523,000
Other products... 22,000
Chase... 15,000
    Total... 1,560,000

Of this $54,000 is figured for wood, which was given to persons holding servitude rights. The expenses were:

Running expenses—
    Central forest bureau $26,000
    Oberfoersters 97,000
    Guards 116,000
    Logging.. 231,000
    Roadmaking.. 47,000
    Planting, sowing, drainage, etc 47,000
    Other running expenses... 128,000
Permanent expenses 60,000
    Total expenses... 752,000
    Real gross income.. 1,522,000
    Net income 770,000

The following figures present the course of these relations for the decade ending 1891:

Financial results for the State forests in Alsace-Lorraine.

Year.	Gross income.	Cut per acre and year.		Per acre of total area.			Price of wood per cubic foot.
		Wood over 3 inches.	Total.	Gross income.	Expenses.	Net income.	
		Cubic feet.	*Cubic feet.*				*Cents.*
1882	$1,337,000	43	55	$3.77	$2.20	$1.55	6.1
1883	1,370,000	42	55	3.86	2.04	1.81	6.5
1884	1,429,000	43	61	4.03	2.04	1.97	6.2
1885	1,301,000	45	59	3.67	2.05	1.61	5.8
1886	1,284,000	45	59	3.62	2.01	1.59	5.7
1887	1,308,000	48	62	3.67	2.06	1.59	5.5
1888	1,335,000	45	57	3.74	1.98	1.74	6.0
1889	1,371,000	46	58	3.84	2.08	1.71	6.2
1890	1,477,000	49	61	4.12	2.06	2.01	6.5
1891	1,522,000	46	56	4.21	2.09	2.12	7.1

The net income, in spite of large yields in wood material and a fairly good market, is comparatively small, though slightly improving. In 1886, when this income was still lower, a special investigation was undertaken, to set forth the reasons of this small net revenue and to suggest improvement. All oberfoersters of note contributed their opinions, and on the whole good results seem to have come from their suggestions for improvement. The chief trouble evidently lies in the great proportion of hardwoods, which leads to a large production of firewood and a small proportion of timber or work wood. Thus 66 per cent of all oak, 91 per cent of all beech, and 83 per cent of all other hardwoods had to be sold as cord and fire wood, bringing generally about 5 cents per cubic foot solid, or about $5 per cord, while for the coniferous woods only 36 per cent has to be sacrificed as cord wood, the rest being sold as timber for just twice the amount obtained for firewood.

This condition of affairs is materially aggravated by the general use of coal as fuel and the rejection of beech as tie timber on railways, etc. This condition of affairs in Alsace-Lorraine is of great interest in considering the forest conditions of the United States. It shows evidently that it is the coniferous timbers which must be looked to as the important ones, and that even large supplies of hardwoods can not be expected to replace such staples as white pine or spruce.

METHODS OF GERMAN FOREST MANAGEMENT.

The following brief description of the methods of German forest management, by which the results described have been attained, was originally prepared in connection with an exhibit at the World's Fair, which the chief of the Division of Forestry collected and installed upon the invitation and at the expense of the German Government, and is mainly reprinted with additions from his annual report for 1893. The description having been based upon the objects exhibited no attempt has been made to alter the form.

MAP WORK AND FOREST DISTRICTING.

The first requirement in the management of any property is that all its conditions should be known and recorded; hence a topographic survey of the forest district to be placed under management is the first requisite. Such survey refers not only to the boundaries and topographical features of the district itself, but also to the surroundings, especially with reference to connections with markets. Finally, for government forests, the geographical position of the forest areas in general should be grouped according to ownership. Maps of the latter description were exhibited from the Governments of Bavaria and of Wurttemberg.

These show in three different colors the forest areas belonging to the Government, to communities and institutions, and to private owners. From these it could be seen not only that the three classes of proprietors share about equally in the ownership of the forest area, but that the Government owns mainly the forests on the mountains, where forest management must be carried on not for profit, but for indirect benefits in the preservation of favorable soil and water conditions, which therefore makes the permanent, well-organized management "by and for the people" necessary. Contrary to the notion to which currency is so often given in the United States, the various governments of Germany do not own more than 35 per cent, exercising partial control (so as to prevent destruction and waste) over only 15 per cent in the hands of communities and institutions, and leaving the balance of 50 per cent of the forest area in private hands almost entirely without restriction.

Sometimes the contours of the country are also indicated on the maps, which serve the useful economic purpose of permitting ready reference of the forest areas to the topography. As an instance of such work there was shown a relief map of Hesse. On this the forest areas were indicated in green color.

For the sake of orderly administration, the whole country is separated into forest divisions or inspections (sometimes both), each of which forms a separate unit of administration.

It is to be understood that we are now speaking only of the Government forests, which are under a uniform general administration.

The administration of the Government forests is usually assigned either to the finance department (as in Bavaria) or to the department of agriculture and forestry (as in Prussia), with one director and council directly in charge under the supervision of the minister or secretary. The position of the director (Oberlandforstmeister) corresponds somewhat to that of our Commissioner of the General Land Office, except that, an extensive technical knowledge being needed in the position, the incumbent is promoted through all positions from the lower grades. Again, each forest division is placed under a separate administrative body consisting of an administrator (Oberforstmeister) with a council of forest inspectors (Forstmeister), each of whom has supervision of a number of the final units of administration, the forest districts (Oberfoersterei, Forstamt). The district officer (Oberfoerster, Revierfoerster, etc.), with a number of assistants, rangers (Foerster), and guards (Schutzbeamte), is then the manager and executive officer in the forest itself, while the higher supervising and inspecting officials are located at the seats of government.

SURVEY OF THE FOREST DISTRICT.

The survey of each forest district is carried out to the utmost minutiæ.

In Prussia the maps of the districts are made on the scale of 1:5,000 in portfolio sheets, representing a careful survey by theodolite of the boundaries of the district, the permanent differences of soil and occupancy (roads, waters, fields, meadows, moors, etc.), and the division of the district into smaller units of management. This kind of map, of which only three copies are made, is then, for purposes of use in daily routine, reduced to a scale of 1:25,000 on one sheet, and printed. The first matter of interest that strikes us on these blank or base maps is the division lines by which the district is divided into parcels or compartments. In the plain these lines divide the district into regular oblong compartments (Jagen) of about 60 to 75 acres each, with sides of 100 and 200 yards, respectively, separated by openings or avenues which we may call "rides" (Gestell, Schneisse), so that the whole makes the appearance very much like the map of an American city regularly divided into blocks. The rides (from 8 to 40 rods wide) running east and west and north and south are lettered, the former, broader ones (main avenues) with capital letters, the latter (side avenues) with small letters, while the compartments are numbered. In the forest itself at each corner a monument of wood or stone indicates the letters of the rides and numbers of the compartments, rendering it easy to find one's way or direct any laborer to any place in the forest. The rides are often used as roads and serve also the purpose of checking fires, etc.

In the hill and mountain districts this regular division becomes impracticable and the lines of compartments conform to the contour, while the opening of the avenues is restricted to those which can be readily transformed into roads; roads, indeed, determining the division lines wherever practicable.

In hill or mountain districts topographic or contour maps become necessary, especially for the purpose of rational road construction, a matter on which in modern times great stress is laid and to which we shall refer later on more in detail. Such contour maps are sometimes executed in papier-maché or gypsum models for readier reference.

PRINCIPLES OF MANAGEMENT.

The fundamental principles upon which the German Government forests and most of the communal and private forests are managed is briefly expressed in the idea that the forest growth is to be treated as a crop to be reproduced as soon as harvested, involving continuity of crops. To carry this principle into effect most advantageously the management must take care to husband the natural forces and conditions upon which thrifty forest growth relies, which leads to the second principle, that of highest efficiency of crops, or the two leading principles combined, to produce the largest amounts of material (or revenue) in the shortest time without impairing the condition and capacity for reproduction of the forest, perpetuating valuable forest growth wherever this is the best crop or where soil conditions make a forest cover desirable. In government forests in addition the financial principle prevails of treating the forest as a permanently invested capital, from which only the interest is to be used, making the amount harvested or the revenue derived to be as nearly alike from year to year or from period to period, and as nearly corresponding to the annual accretion, as it is possible to make them.

The present Oberlandforstmeister, or director, of the Prussian forest department uses the following language in laying down the principles upon which the Government manages its forests:

The Prussian State forest administration does not accede to the principles of a continuous highest soil rent based upon compound interest calculations, but believes, in contradistinction to private forest management, that it can not avoid the obligation in the management of the State forests of keeping in view the welfare of the whole community of citizens, and therein taking into consideration the need for continued supply of wood and other forest products as well as the other objects to which in so many directions the forest is subservient. The administration does not consider itself entitled to pursue a one-sided financial policy, least of all to submit the Government forests to a pure money-making management strictly based on capital and interest calculations, but considers it its duty to so manage the forests as a patrimony belonging to the whole nation that the present generation may be benefited by the highest possible usufruct in satisfying its wants and deriving the protection which the forest renders, and that to future generations may be secured at least as large usufruct of the same kind.

To carry out these principles the intimate knowledge of the conditions of the property, referred to above, is necessary and is obtained by a careful forest survey as a basis for a systematic administration and forest regulation.

These data of this forest survey are not only recorded in writing but such as can be readily noted are finally placed upon the blank maps described above, together with the results of the forest regulation described further on, so that the manager can readily overlook the details of his district and the propositions for its management.[1] This information—after further subdivision of the compartments where needed on account of differences in soil conditions or growth—is given by means of different colors, difference in shade, numbers, figures, marks, and signs. These maps, which are prepared after a most painstaking forest survey, and which we may call "manager's map" (Plate XXXII), show at a glance not only the nature of soil conditions and what the principal kind of timber and its admixtures are in each compartment or subdivision, but also how old the growth; whether it is to be treated as a coppice, standard coppice, or timber forest; at what period in the rotation it is to be cut, and such notes as the manager himself may add from year to year, as, for instance, the yearly fellings, plantings, movable nurseries, new road projects, etc.

One of the most instructive exhibits in this direction was that showing the changes in Timlitz forest, Saxony. The map of the district in 1822 presented about the condition of one of our mismanaged Michigan forests of pine and hard woods mixed, from which all the good timber had been culled, leaving it to inferior kinds with few groups of straggling pines and more valuable hard woods, without symmetry or system in the distribution of kinds or age classes. At the same time a map was constructed showing ideally how the forest was to look after eighty years' well-planned management. We can then follow in the maps made every ten or twenty years the changes in appearance under the hand of the forester. During the management new information and experience have dictated modifications of the original working plan, giving rise to a new manager's map, the approach to which appearing in the timber map for 1885 leaves no doubt that at the end of the period of regulation we will have a well-grown pine forest, with deciduous trees mixed in or confined to the more suitable situations, so disposed over the area that annually or periodically the same amount, or nearly so, of valuable material can be harvested.

The painstaking methods of surveying, describing, measuring, calculating, planning, book-keeping, and repeated revising of all the work from decade to decade were shown in the regulation work of the district of Hinternah, Prussia, contained in six large folio volumes of manuscript, continued from the year 1822 to the last revision in 1890. We can only briefly indicate what this work involves, which was briefly summarized in the following exhibit:

FOREST REGULATION.

PROGRESS OF WORK REQUIRED TO BRING FOREST AREAS UNDER RATIONAL FOREST MANAGEMENT.

I. Geodetic and topographic survey and mapping.

II. Forest survey in connection with I, noting all areas distinguished by quality of soil, composition, and age of timber; general description of forest conditions, of climatic conditions, of surrounding conditions, of possible dangers, of market conditions, means of transportation, etc.

[1] Each State government pursues somewhat different methods of mapping. Sometimes two sets of maps are made, one to show the conditions, which might then be called a timber map, the other to show the working plan; but these are now mostly combined into one.

III. *Forest districting.* Division of forest into parcels or lots and aggregation of lots into blocks and ranges. In the plain, rectangular lots, divided by cleared lines called rides (Gestell), are customary; in hilly and mountainous country division lines follow the configuration of soil. Differences of soil or character of growth within lots give rise to formation of sublots.

IV. *Forest yield valuation* (assessment). Ascertaining amounts of timber standing, rate of growth on various sites, determining capability of production and future yield in material and money.

V. *Determining plan of management* (working plans). General plan for all time; special plans for period of ten to twenty years. Determining length of rotation; amounts annually to be cut, designating lots to be cut, with a view to obtaining favorable distribution of age classes; thinnings to be made; methods to be used in felling and cultures.

METHODS OF FOREST REGULATION.

In Prussia it was Frederick the Great who first ordered a regulated administration of the Government forests soon after the beginning of his reign. The first simple prescriptions of dividing the forests into equal areas and cutting every year a proportionate area were followed up with more elaborate ordinances, having in view a closer equalization of the amounts of material harvested and revenues obtained, besides other considerations of management for continuity, until finally the basis for present methods of regulation was reached in the ordinance of 1836, since modified in its details, under which "the preservation, revision, and perfection of the work of forest valuation and regulation" is carried on.

The modus operandi, similar in principle in all Government forest administrations, is about as follows:

Let us assume that the Government has purchased [1] a new forest district, comprising, say, 10,000 acres, the average size of the existing districts. The necessary surveys and blank maps, as explained, have been made and the boundaries carefully established in the field, the division into compartments or parcels, larger or smaller according to the need of a more or less intensive management, have been noted on the maps and marked on the ground (the avenues perhaps partially opened), and for the sake of satisfactory administration a number of the parcels have been combined into subdistricts, "blocks," or ranges; and thus the first—purely geometrical—basis for a rational administration has been established. Now the arithmetical basis is to be ascertained. For this, in the first place, a general description of the district in its present condition is desirable, parts of which, however, can be furnished only after the more thorough measurements described later. Such a description recites all needful knowledge regarding the extent, the manner of division, the boundaries, and the legal rights. Next follows a description in general terms of topography, climate, and soil conditions, and of the forest growth, being a condensation of the special description by parcels. The manner of treatment hitherto, the market conditions, current market prices, and usual wages are noted. Then, after recital of the processes and methods by which the information in the following detail work has been obtained, the principles adopted for the management and its motivation are stated, forming a general guide for the manager for all time.

These principles are formulated by a commission after sufficient general knowledge of the condition of the district is obtained. In this important part of the general description not only the territorial partition of the district into compartments and blocks or ranges is determined, and reasons given for it, but also the system of management for each block or parts of blocks, whether

---

[1] Prices for forest soil vary, of course, according to their location and condition, just as in our country. In 1819 Bavaria sold 27,000 acres of her State forests at $68 per acre. In Prussia the Government has lately (1884–1887) paid prices ranging from $5 to $60 per acre, and for a round 70,000 acres the price per acre was $24 average. These were mostly devastated waste lands in the northern plain. In Thuringia, where prices for wood and land are higher, the price for forest land is from $20 to $60 and as high as $80. These prices do not, of course, include any timber growth, the value of which, if present, is calculated according to well-known careful methods of determining "expectation values." According to a calculation by Dr. J. Lehr, based on the net income as representing interest at a 3 per cent rate, and assuming a ninety-year rotation of the forest growth for the entire German Empire, the forest land was worth $25 per acre and the wood on it $156 per acre.

coppice, standard coppice, timber forest, etc.,[1] and the length of rotation—i. e., the time within which a block is to be cut over and reproduced; furthermore, the principles according to which the fellings are to progress, reproduction is to be secured, thinnings are to be made, the annual yield to be expected, and the time within which the forest is to be brought into a regular systematic order of management—in short, all the general framework of the management as far as determining a set policy into which the special working plans should fit. Before this report can be made final, however, the work of the valuator or examiner must have proceeded to some extent.

VALUATION WORK.

The valuator or estimator, upon whose work as a basis the general and special working plans depend, begins by examining and describing briefly the conditions of the soil, its productive capacity, and the kind and appearance of the growth in each compartment (or subparcel, if conditions of growth or soil make such subdivision desirable). In the description the dominating kind of timber, or, if mixed in equal proportions, that upon which the management is to be promi-

[1] NOTE.—Timber forest (Hochwald, high forest) is a forest in which trees are allowed to grow to maturity, and reproduction is effected either by natural seeding from the old growth in various ways, or by planting or sowing after removal of the old growth; it is usually managed in rotations of 70 to 120 years.

Coppice (Niederwald, low forest) is a forest in which reproduction is expected by sprouts from the stumps; this is usually managed in rotations of 10 to 40 years.

Standard coppice (Mittelwald, middle forest) is a combination of the two former, the standards being allowed to grow to maturity and reproduction being secured both by seed and sprouting.

Determining the rotation.—Our friends who are attempting to bring about a more rational treatment of our forests have often a mistaken notion as to when timber should be cut, when it is ready for the harvest. This can not be determined by any set period, as in the ripening of fruit in agriculture, or by any more or less defined age, much less by any diameter measure. The determination of the "felling age" (Haubarkeitsalter) or of the length of "rotation" (Umtrieb) depends on the use to which the crop is to be put, the manner in which it is to be reproduced, and the amount of material that can be produced, or the amount of profit that can be derived from it. This determination is one of the most difficult, requiring both careful financial calculation and knowledge of forest technique.

The "silvicultural rotation" is that which considers mainly the forest technique, being the time when perfect natural reproduction is most surely attainable—i. e., fullest seed production in timber forest, highest sprouting capacity in coppice forest; or when preservation of the productive capacity of the soil, avoidance of damage from windfalls, diseases, etc., are uppermost considerations. These considerations of course also influence in part the determination of any of the following rotations, which we may call "economic rotations."

The "rotation of greatest material production" is that which allows the forest to grow as long as the average annual accretion is at a maximum. This differs of course with species, climate, soil, etc. If for the mass of material we substitute its money value and strive to so arrange that the time of rotation coincides with the largest money returns, we have a "financial rotation."

Various points of view lead to different kinds of financial rotations:

"Rotation of the highest harvest value," or "technical rotation," which attempts to produce certain desired sizes and qualities in largest quantity with a view of obtaining thereby the largest money return for the crop under the circumstances (management for telegraph poles, fence posts, osier holts, tan-oak coppice).

"Rotation of the highest forest revenue," when the growth is to be harvested at the time of its maximum average annual net money value; this time is influenced both by the amount of material and the price paid for better sizes and quality of wood. In this rotation no regard is paid to the original capital invested in the soil; when this latter factor is introduced into the calculation we arrive at the true "financial rotation" or "rotation of the highest soil (or ground) rent," in which the forest is to be cut at a time when the capital invested in soil, stock, and management furnishes the highest interest rate. This capital, as far as the soil is concerned, may be represented by its actual cost or by its market value, or else by its capacity for production (Bodenerwartungswerth; soil-expectation value), which is found by adding the values of expected returns at harvest discounted to the present time and deducting the expenses incurred up to the time of harvest, similarly discounted.

To determine this value experience tables must give the data. Local conditions and prices and the rate of interest applied of course influence the length of the financial rotation. It is shortest for a firewood management (in Germany, say 60 to 70 years), for spruce and pine at an interest rate of 2 to 3 per cent a rotation of 70 to 90 years, with oak 120 years, appear as profitable rotations; where small sizes, mining timber, posts, poles, etc., are bringing good prices, the most profitable financial rotation may be shorter. It stands to reason that the length of this rotation, as well as of all others, can be only approximately calculated. The forestry literature of Germany is most prolific just now with regard to determining financial rotations, and the highest mathematical skill is employed in the discussion.

Growth (Bestand, stand) is here and further on used in the collective sense of the word to denote an aggregate of trees, for which also the word "stand" may be employed.

nently based, is named first, and the average age of the growth with special reference to the dominating timber is ascertained for the purpose of ranging the parcel into an "age class," which comprises usually twenty years, so that the growth of 1 to 20, 21 to 40, 41 to 60 years, etc., form each an age class or period. The density of the growth and larger openings devoid of tree growth are specially noted. The valuator at the same time is expected to form, from general appearances, an opinion as to the best treatment of each parcel in the near future, and note it, and especially whether the growth is to be cut during an earlier or later period than its age would warrant, considering the likelihood of its thrifty or its unsatisfactory growth. He also estimates the amounts to be taken out in thinnings for the next twenty years.

With this information established a table may be constructed, in which the area of each parcel is entered, according to its average age or "age class," modified by considerations of productive capacity, and from this a "timber map" is made, showing the present conditions of the forest, the kind of dominating timber in each parcel being denoted by a color, intermixed timbers by signs, and the age by the shade of the color in 4, 5, or 6 gradations, according to the number of age classes, as shown in the accompanying ideal map.

ARRANGEMENT OF AGE CLASSES.

Now follows the determination of the future arrangement of age classes, the object of which is to have, when the forest is regulated, in each period of the rotation an approximately equal or equally producing area to be cut. It therefore becomes necessary to shift the distribution of age classes, in order to attain the equality of the sum of areas in each period. In addition to the mere equalization of areas, there are several other considerations guiding the valuator in arranging the age classes. The oldest timber, as well as that which for some reason has ceased to make satisfactory growth, is of course to be cut first; hence the conditions of these areas are more specially examined regarding health, density of cover, soil, vigor, etc. In coniferous growths, especially in the plain, the danger from windfalls, if one parcel is cut and thereby the other exposed to the prevailing storms, necessitates such an arrangement in the location of the fellings (or age classes) that the removal of an old growth will leave behind it a young growth which is less liable to be thrown. This local distribution of the age classes by which, in the direction of the prevailing winds, no two neighboring growths are assigned to the same period is also desirable from other considerations. By avoiding a series of extensive fellings side by side the danger from fires is lessened and liability to spread of diseases and insect attacks, danger from frost, and drought to young growths is confined or reduced. Hence an arrangement of the age classes as near as possible after the following scheme has been generally adopted, in which the Roman figures denote the age classes, I standing for the oldest growth, containing, if the rotation has been set at 100 years, timber of 80 to 100 years, to be felled within the first twenty years; II for that to be felled within twenty-one to forty years from the present, and so on; V to be felled in from eighty to one hundred years.

V	III	I	IV
IV	II	V	III
III	I	IV	II
II	V	III	I

Prevailing winds⟶

Fig. 24.   Diagram showing arrangement of age classes.

In mountainous districts, where the topography influences the expense of transportation, fellings are often more concentrated and the higher parcels used and reproduced before the lower, in order to avoid injury to the young growth by a reversed condition when the material from above would have to pass through the young growth below. Various minor points may also dictate exceptional arrangement. In coppice growth, needed protection of the stocks against cold north winds makes it desirable to have the fellings progress from the south and west toward north and east. Altogether it will have become apparent that the distribution of successive fellings is an important matter, not only from the standpoint of regulated administration, but also of successful culture.

In the accompanying map (Pl. XXXII) we have attempted to give an idea of the matter on which a "manager's map" is constructed, and how ideally in a forest of the plain the arrangement of age classes would appear when the forest regulation is perfected.

YIELD CALCULATIONS.

When the distribution of areas has been effected in accordance with the considerations set forth, the yield calculations are made. These are computed after careful measurements and by various methods of calculation, which have been developed after much experience during more than one hundred years.

Since the different compartments are cut at different times, not only the present "stock on

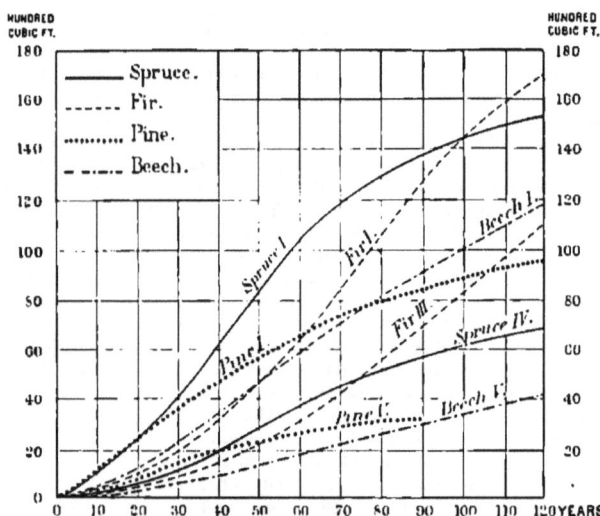

Fig. 24.—Diagram showing comparative progress of yields of spruce, fir, pine, and beech on best and poorest site classes.

hand" needs to be measured, but also the accretion for each age class from the present to the middle of the period in which it is to be utilized as to total quantity (decreasing in arithmetical proportion as the stock on hand is diminished by fellings), when by adding the two quantities and dividing the total by the number of years in the rotation or time of regulation the equalized yearly quota to be utilized, or "felling budget" (Haubarkeitsertrag or etat), can be calculated.

The determination of existing stock is made by measuring diameter breast high by means of calipers, estimating the average height, and calculating contents with the aid of tables which give the corresponding volumes of timber wood (above 3 inches diameter). These tables are constructed after numberless detail measurements, from which the "factor of shape" for each species, soil, or climate is derived, for, since the tree is neither a cylinder nor a cone, which could be calculated from the base and height, the modification from either of these two forms, the "factor of shape" must be determined experimentally in order to arrive at the approximately true contents. In very irregular growths and with skillful valuators a simple estimating of contents or the use of so-called normal yield or "experience tables," which give for the various species, soils, and climates the amount of wood that would normally be produced per acre at a given period, is not excluded.

Normal yield table for spruce.

[Main growth (exclusive of thinnings) per acre]

Age.	Number of trees.	Cross-section area of all trees breast high.	Average height.	Wood above 3 inches diameter.	Wood, total mass.	Age.	Number of trees.	Cross-section area of all trees breast high.	Average height.	Wood above 3 inches diameter.	Wood, total mass.
Site class I.						*Site class III.*					
		Sq. ft.	*Feet.*	*Cu. ft.*	*Cu. ft.*			*Sq. ft.*	*Feet.*	*Cu. ft.*	*Cu. ft.*
10 years		49.2	4.9	86	415	10 years		18.3	1.9		200
20 years	2,591	111.4	16.7	1,101	2,174	20 years		55.7	6.6	100	772
30 years	1,700	150.5	29.2	2,603	4,204	30 years	3,732	86.6	15.7	472	1,617
40 years	1,065	188.1	47.6	4,748	6,378	40 years	2,412	130.1	25.6	1,244	2,760
50 years	721	209.7	62.6	7,222	8,623	50 years	1,580	154.9	36.7	2,574	4,247
60 years	515	225.6	76.7	9,299	10,625	60 years	1,056	171.8	48.2	4,091	5,634
70 years	390	237.1	88.2	10,582	12,198	70 years	734	185.3	59.0	5,249	6,893
80 years	321	244.9	97.4	11,655	13,213	80 years	500	195.2	67.9	6,220	7,991
90 years	269	250.9	105.3	12,555	14,043	90 years	424	205.2	74.1	7,093	8,866
100 years	243	258.4	112.5	13,299	14,715	100 years	380	214.9	79.4	7,922	9,638
110 years	229	264.5	117.7	13,971	15,272	110 years	316	223.2	88.0	8,691	10,296
120 years	226	269.7	121.4	14,586	15,730	120 years	320	230.6	85.6	9,324	10,725
Site class II.						*Site class IV.*					
10 years		26.1	3.2		415	10 years		11.3	1.6		157
20 years		77.9	11.5	315	1,201	20 years		36.5	4.6		500
30 years	2,361	89.9	22.6	1,187	2,460	30 years		72.2	10.5	140	1,044
40 years	1,619	151.8	35.1	2,562	4,016	40 years	3,164	107.9	18.0	515	1,830
50 years	1,161	160.1	47.2	4,176	5,791	50 years	1,968	130.1	26.2	1,287	2,708
60 years	842	200.1	59.7	6,239	7,851	60 years	1,276	143.5	35.1	2,291	3,761
70 years	639	213.6	71.8	7,806	9,481	70 years	864	154.9	42.0	3,089	4,510
80 years	481	222.7	83.0	9,295	10,725	80 years	648	162.6	51.5	3,780	5,218
90 years	350	231.3	91.5	10,309	11,683	90 years	554	172.3	57.1	4,361	5,763
100 years	301	239.2	97.7	11,125	12,394	100 years	500	181.5	61.3	4,818	6,299
110 years	293	246.5	101.0	11,740	13,013	110 years	464	187.0	63.3	5,305	6,707
120 years	291	272.3	106.6	12,293	13,585	120 years		191.4	66.6	5,720	7,150

In very regular growths trial areas only are measured. The more usual manner of determining the rate of accretion, however, for purposes of yield calculation, is by felling sample trees of each class, dissecting and measuring the accretions of past periods.

In modern times the exact measurements are mostly confined to the growths that are utilized during the first or first two periods of twenty years.

FELLING BUDGET.

After all these data for each compartment have been booked, and the yield of branchwood and roots—for even these are mostly utilized—as well as the probable amounts to be taken out in thinnings, have been estimated and recorded, and after the likelihood of decreased accretion in the different compartments has also been determined from measurements and experience, the "felling budget" is determined as a sum of the stock on hand and the amount of annual accretion multiplied by the time, during which it is allowed to grow, i. e., in the average to the middle of the period in which the compartment is placed, divided by the period of rotation. Thus a growth of eighty-five years, which showed a stock on hand of 3,825 cubic feet per acre, and hence had an average accretion hitherto of 3,825 ÷ 85 = 45 cubic feet per year, which is likely to be reduced on account of gradual reduction in stock and other untoward conditions to 30 cubic feet, would yield during the first period 3,825 + 30 × 10 = 4,125 cubic feet. And if the compartment contained 50 acres it should be credited in the working plan in the column for the period 1 with 4,125 × 50 = 206,250 cubic feet. By adding up the amounts of the yield of all the compartments placed in the first period and dividing by 20 (the length of the period) the annual budget which should be felled during the period is found. If, however, it is desired to equalize the fellings more or less through a longer period— for instance, the time of rotation—then the amounts in all the periods must be summed up, and these sums as nearly as possible equalized by shifting the position of the compartments from one period into another (necessitating always new calculations of the accretion) until the equalization in the periodic sums is effected.

Even then, however, before finally determining the annual budget, a calculation is made to see whether the area contains as much timber as it normally should; if more, the budget may be increased; if less, a saving must be made in order to bring up the stock on hand to the normal. If, for instance, we know from the experience tables that our forest should normally yield 50 cubic feet per acre a year in a 100-year rotation, then the normal stock would be 100 × 50 ÷ 2 = 2,500 cubic feet per acre. This is the average amount of wood per acre which we should strive to keep in stock in order to get the full benefit of the productive capacity of the soil and insure an equal growth and equal annual cut for all time. In reality this ideal is, of course, never reached, but this so-called normal forest, conceived in ideal condition, serves as a guide in the working plans, and the conception is a most useful and important one. To put it into practice we must either save at first on the annual cut until normal condition is attained, or we may increase the cut if more old timber than necessary for normal stock is on the ground. Additional reserves may also be provided for to avoid any unforeseen shortcomings in the budget due to insect ravages, mistakes in calculations, etc.

We can not here enter into the details of all the work of the valuator, being satisfied with having indicated in general the methods pursued. In coppice management, of course, all these fine calculations become unnecessary, and the periodical or annual cut is determined by area mainly.

From the general plan thus elaborated the special plan for the first period or half period of the management is worked out in detail both for fellings, cultures, and other work, road building, drainage, etc. This special plan, then, is the basis on which the local manager finally makes out the annual plans of work, which are submitted for revision and approval to the controlling officers. Thus, while the general and special working plans lay down the general principles, the annual plans, into which enter considerations of immediate needs and financial adjustments, permit such deviations from the general plans as may appear needful from year to year. Every ten or twelve years, or at other stated periods, a careful revision of the whole regulation work is made, in which the carefully noted experiences of the manager are utilized to correct and perfect the plans.

FOREST PROTECTION.

In this country the greatest danger to the forest, besides the indiscriminate cutting, is to be found in fires. How little this scourge of American forests is known in Germany may appear from the statistics of fires in the Government forests of Prussia (representing 60 per cent of the German forest area), 56 per cent of which are coniferous, which show that railroading may be carried on without the necessity of extra risks, if proper precautions are provided. During the years 1882–1891 there had occurred 156 larger conflagrations—96 from negligence, 53 from ill will, 3 from lightning, and only 4 from locomotives. Seven years out of ten are without any record of fire due to this last cause.

From 1884 to 1887 fires occurred in Prussia on 3,100 acres, but only 1,450 were wholly destroyed, i. e., 380 acres per year, or 0.005 per cent of the total area of Government forests. In Bavaria during the years 1877–1881 only 0.007 per cent of the forest area was damaged by fire, and the loss represented only 0.02 per cent of the forest revenues. During the unusually hot and dry summer of 1892 only 19 fires, damaging more or less 5,000 acres, occurred.

Besides the thorough police organization and the compartment system, which permits not only ready patrolling but also ready control of any fire, the system of safety strips, described in the report of this division for 1892, where a fuller discussion of this subject may be found, prevents the spread of fire from locomotives.

A much more fruitful cause of damage to the cultivated forests of Germany is found in insect ravages. The annual expenditures in fighting and preventing these in the Prussian Government forests in ordinary times amount to about $50,000. Caterpillars and beetles eat the leaves, and thereby reduce the amount of wood produced and the vitality of the tree; bark beetles follow and kill it; borers of all kinds injure the timber. Hence entomology, the study of life habits of the injurious insects and the methods of checking their increase, forms part of the forester's work.

Fungus growth and decay kill the standing tree and injure the cut timber. The study and methods of counteracting this injury form, therefore, part of the work of the forester.

FOREST CROP PRODUCTION OR SILVICULTURE.

While we have so far considered mainly the administrative and managerial features of German forestry practice, we come now to the most important and truly technical branch of the art, namely, the forest crop production or forest culture. This part we may call forestry proper, for while the methods of forest regulation, forest utilization, and forest protection, which may be comprised in the one name, "forest economics," are incidental, and may differ even in principle in various countries and conditions, the methods of crop production or forest culture, being based on the natural laws of the interrelations of plants to soil and climate, must, at least in principle, be alike all over the world. Here pure forestry science finds its application and development.

These principles have been elucidated more fully in the next chapter. We will, therefore, here only briefly restate the more important ones with some of their applications in German practice.

PLANTING.

Seemingly the simplest and easiest way of reproducing the crop is that practiced in agriculture, namely, removing the entire mature crop and sowing or planting a new crop. But this method, which has been so largely practiced in Europe and admired by our countrymen and writers on forestry, has its great drawbacks, which have of late become more and more apparent, and the tendency now is to return more and more to the "natural reproduction." While the simplicity of the method of clearing and planting recommends itself for a routine or stereotype management, it has not always proved as successful as would be expected. The large clearings which the young planted seedlings are unable to protect from the drying influences of sun and

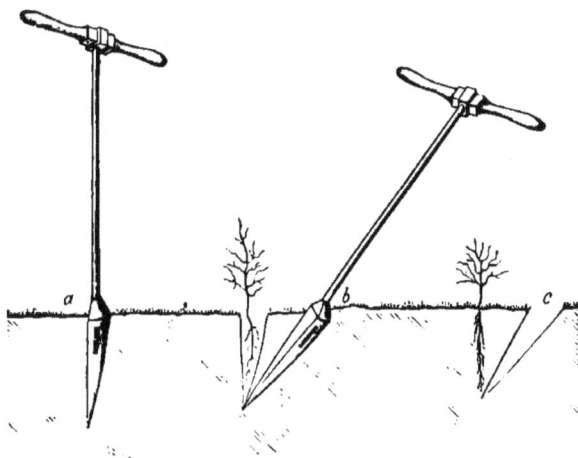

Fig. 25.—Iron dibble used in setting out small pine seedlings.

wind bring about a desiccation and deterioration of the forest soil and an enormous increase of insect pests, while other dangers in later life from wind and disease have been largely the result of these uniform growths. And when it is understood that to secure a desirable stand the plantings must be gone over and fail places replanted five, six, and more times, it becomes apparent that the method is extremely expensive, and hence the proper treatment of the natural crop with a view to its reproduction by natural seeding is the most important part of forest culture. Yet under certain conditions, and where no natural crop to manage is found, planting or sowing becomes a necessity, and various methods and tools have been developed to meet various conditions.

It would exceed the limits of this report to describe these various methods; we can refer to only one of the simplest and cheapest with which every year many millions of small 1 or 2 year old pine seedlings are set out in soils which do not need or do not admit of preparation by plow or spade. The instrument used is an iron dibble (fig. 25); the shoe, with one rounded and one flat side, in shape like a half cone, 8 inches long with 3½-inch base; the handle, a five-eighths-inch rod, 3½ feet long, is screwed into the base of the shoe and carries a wooden crossbar, by which the instrument is handled. The modus operandi is to thrust this iron dibble into the ground; then by moving it lightly back and forth to somewhat enlarge the hole and withdraw it; a boy or girl puts the plantlet in the hole to the flat side; the dibble is thrust again into the ground 1 to 1¼ inches back of the first hole somewhat slantingly toward the bottom, and pressed forward to fasten the plant in its stand; then by irregular thrusts the last-made hole is obliterated. Two planters with a boy, carrying the plants in a mixture of loam and water to keep the roots moist and also heavy for better dropping, may set 5,000 plants in a day.

INTRODUCTION OF EXOTICS—WHITE PINE YIELDS.

The valuable species of trees indigenous to Germany which are subject to special consideration in forest management are but few. The most important forest-forming ones are 1 pine, 1 spruce, 1 fir, 1 larch, 1 oak, 1 beech, 1 alder. In addition we find of broad-leaved trees a blue beech, 1 ash, 3 kinds each of elm, maple, and poplar, in some parts a chestnut, and 2 kinds of birch and linden, and several willows, together with some 8 or 10 kinds of minor importance, while of conifers in certain regions 4 other species of pines are found. Some years ago the attention of European foresters was forcibly turned to the richness of the American forest flora, and a movement set in to introduce exotic tree species which might be more productive or show better qualities than the native. Our white pine, a good-sized section of which was exhibited, had been quite extensively planted in the beginning of this century, and these plantations, some 80 or 90 years old, are now coming into use. The quality of the wood, however, has not as yet found much favor, but the quantity per acre exceeds that of any of the native species. Records are extant which show, at 70 years of age, a yield of 14,000 cubic feet of wood containing about 70,000 feet of lumber B. M. per acre.

On moderately good forest soil in Saxony a stand 78 years old contained over 400 trees per acre, of which three-fourths were white pine, the rest spruce, larch, beech, and oak. Only 5 white pine trees were under 70 feet high, the majority over 80. Notwithstanding the crowded position, only 15 trees were under 8 inches diameter, the majority over 12 inches, the best 28 inches. The total yield was 12,880 cubic feet of wood per acre, besides the proceeds of previous thinnings. The rate of annual accretion in cubic feet of wood for white pine in the last years amounted to 2.5 per cent of the total contents of the trees, or about 0.4 cubic foot per tree. Of the trunk wood at least 90 per cent could be utilized for lumber, since the shape of these trunks was so nearly cylindrical as to be equal in contents to one-half a perfect cylinder of the height and diameter of the trees taken breast high.

A stand 82 years old on poor land produced 12,500 cubic feet of wood, indicating an average yield for the eighty-two years of 212 cubic feet of wood per annum, of which about 700 feet of lumber B. M. could be calculated. On very poor soil and planted very thick without admixture of hard woods it produced trees 24 feet high and 5 inches thick in twenty years; and on fairly good soil trees 54 feet high, 11½ inches thick, in thirty to thirty-five years, excelling in either case the native spruce (*P. excelsa*) both in height and thickness.

It is also of interest to mention in this connection that a plantation of about 7 acres in the city forest of Frankfort-on-the-Main during the eighteen years ending 1884 brought $115 rent per year for the privilege of seed collecting alone; failing to produce seed only three out of the eighteen years and yielding a maximum of $500 rent during one of the eighteen years; much of the seed finding a market in the United States.

Besides the white pine, the black locust has also for quite a long time found a home in the plantations of Europe, but the species which are now propagated in large quantities, having after trial shown superior advantages in behavior and growth, are our Pacific coast conifers, the Sitka spruce, the Douglas spruce, the Lawsons cypress, and the Port Orford cedar, sections and photographs of which, grown in Germany, were exhibited, as well as of black walnut and hickory. These trees are now used to plant into fail places or openings, in groups or single individuals, and are especially prized for their soil-improving qualities and their rapid growth.

The methods of management for natural reproduction are generally divided into three classes, namely, the coppice, when reproduction is expected from the stumps; the standard coppice, when part of the growth consists of sprouts from the stump and another part of seedling trees; and the timber or high forest, when trees are grown to maturity and, unless harvested and replanted, reproduction is effected entirely by natural sowing.

COPPICE MANAGEMENT.

This practice is employed for the production of firewood, tanbark, charcoal, and wood of small dimensions, and is mostly applicable only to deciduous trees. The capacity of reproduction from the stump is possessed by different species in different degrees, and depends also on climate and soil; shallow soil produces weaker but more numerous shoots than a deep, rich soil, and a mild climate is most favorable to a continuance of the reproductive power. With most trees this capacity decreases after the period of greatest height-growth; they should therefore be cut before the thirtieth year, in order not to exhaust the stock too much. The oak coppices for tan bark are managed in a rotation of from ten to twenty years. Regard to the preservation of reproductivity makes it necessary to avoid cutting during heavy frost, to make a smooth cut without severing the bark from the stem, and to make it as low as possible, thus reducing liability to injuries of the stump and inducing the formation of independent roots by the sprouts.

It will be found often that on poor and shallow soil trees will cease to thrive, their tops dying. In such cases it is a wise policy to cut them down, thus getting new, thrifty shoots, for which the larger root system of the old tree can more readily provide. This practice may also be resorted to in order to get a quick, straight growth, as sprouts grow more rapidly than seedlings, the increased proportion of root to the part above ground giving more favorable conditions of food supply. It must not be forgotten, however, that this advantage has to be compensated somewhere else by a disadvantage; sprouts, though growing fast in their youth, cease to grow in height at a comparatively early period, and for the production of long timber such practice would be detrimental.

Regard to the preservation of favorable soil conditions, which suffer by oft-repeated clearing, requires the planting of new stocks where old ones have failed. Mixed growth, as everywhere, gives the best result. Oaks, walnut, hickory, chestnut, elm, maples, birch, cherry, linden, catalpa, and the locust also, with its root-sprouting habit, can be used for such purpose.

If when cutting off the sprouts, at the age of from 10 to 20 years, some trees are left to grow to larger size, thus combining the coppice with timber forest, a management results which the Germans call "Mittelwald," and which we may call standard coppice management.

STANDARD COPPICE.

This is the method of management which in our country deserves most attention by farmers, especially in the Western prairie States, where the production of firewood and timber of small dimensions is of first importance, while the timber forest, for the production of larger and stronger timbers, can alone satisfy the lumber market. The advantages of this method of management, combining those of the coppice and of the timber forest, are:

(1) A larger yield of wood per acre in a short time.
(2) A better quality of wood.
(3) A production of wood of valuable and various dimensions in the shortest time with hardly any additional cost.
(4) The possibility of giving closer attention to the growth and requirements of single individuals and of each species.
(5) A ready and certain reproduction.
(6) The possibility of collecting or using for reforestation, in addition to the coppice stocks, the seeds of the standards.

The objections to this mode of treatment are the production of branches on the standards when freed from surrounding growth, and the fact that the standards act more or less injuriously on the underwood which they overtop.

The first objection can be overcome to a certain extent by pruning, and the second by proper selection and adjustment of coppice wood and standards. The selection of standards—which preferably should be seedlings, as coppice shoots are more likely to deteriorate in later life—must be not only from such species as by isolation will grow into more useful timber, but if possible from those which have thin foliage, thus causing the least injury by their cover to the underwood. The latter should, of course, be taken from those kinds that will best endure shade. Oaks, ashes, maples, locust, honey locust, larch, bald cypress, a few birches, and perhaps an occasional aspen, answer well for the standards; the selection for such should naturally be from the best-grown straight trees. The number of standards to be held over for timber depends upon the species and upon the amount of undergrowth which the forester desires to secure. The shadier and the more numerous the standards the more will the growth of the coppice be suppressed. From a first plantation one would naturally be inclined to reserve and hold over all the well-grown valuable saplings. The coppice is, of course, treated as described above.

As before mentioned, on account of the free enjoyment of light which the standards have they not only develop larger diameters, but also furnish quicker-grown wood (which in deciduous trees is usually the best) and bear seed earlier, by which the reproduction of the forest from the stump is supplemented and assisted. Any failing plantation of mixed growth, consisting of trees capable of reproduction by coppice, may be recuperated by cutting the larger part back to the stump and reserving only the most promising trees for standards.

If equally well-grown coppice and standards are desired, a regular distribution of the standards, mostly of the light-needing, thin-foliaged kinds, should be made. If prominence is given to the production of useful sizes, the standards may be held over in groups and in regularly distributed specimens, in which case those of the shade-enduring kinds are best in groups.

THE TIMBER FOREST.

In the timber-forest management we may note various methods: The method of selection (Plenterwald), in accordance with which only trees of certain size are cut throughout the whole forest, and the openings are expected to fill up with an after-growth sown by the remaining trees. This method prevailed in former ages, but was finally almost everywhere abandoned because of the difficulty of organized administration and control of such an irregular forest containing trees of all ages, and because the after-growth is apt to progress but slowly with fore-grown trees surrounding and overshadowing it, or may consist of worthless kinds. Of late a revival of this method with various modifications designed to meet the objections is noticeable; the advantage of keeping the soil constantly shaded and thereby preserving the soil moisture also recommending this method. More uniform growths, more regular distribution of age classes, and a more regulated administration was possible by various "regeneration methods," by which a certain area—a compartment—would be taken in hand and the cutting so systematically directed that not only a uniform young growth would spring up through the whole compartment, but by the gradual removal of the mother trees light would be given to the young growth as needed for its best development. This method (Femelschlag) is practiced almost exclusively in the extensive beech forests, somewhat in the following manner:

REGENERATION METHODS.

In the first place it is necessary to know the period at which a full seed year may be expected. This differs according to locality and kind. One or more years before such a seed year is expected the hitherto dense crown cover is broken by a preparatory cutting of the inferior timber, enough being taken out to let in some light, or rather warm sunshine, which favors a fuller development of seed, the increased circulation of air and light at the same time hastening the decomposition of the leaf-mold and thus forming an acceptable seed bed.

As soon as the seed has dropped to the soil, and perhaps, in the case of acorns and nuts, been covered by allowing pigs to run where it has fallen, a second cutting takes place uniformly over the area to be regenerated, in order that the seeds may have the best chance for germination—air, moisture, and heat to some degree being necessary—and that the seedlings may have a proper enjoyment of light for their best development and yet not be exposed too much to the hot rays of the sun, which, by producing too rapid evaporation and drying up the needful soil moisture, would endanger the tender seedlings. This cutting requires the nicest adjustment, according to the state of the soil, climatic conditions, and the requirements of seedlings of different kinds.

While the beech requires the darkest shade, the pine tribe and the oaks demand more light, and should, by the successive cuttings, be early freed from the shade of the mother trees. Beech seedlings are more tender, and only by the gradual removal (often protracted through many years) of the shelter of the parent trees can they be accustomed to shift for themselves without liability of being killed by frost. The final cutting of the former generation of trees leaves many thousand little seedlings closely covering the soil with a dense shade.

That the method of management must differ according to species and local conditions is evident; and in a mixed forest especially are the best skill and judgment of the forester required to insure favorable conditions for each kind to be reproduced. It is to be expected that such seedlings are rarely satisfactory over the whole area, and that bare places of too large extent must be artificially sown or planted.

Another method is the "management in echelons" (Coulissen, Saumschlag), which consists in making the clearings in strips, and awaiting the seeding of the clearing from the neighboring growth. It is applicable to species with light seeds, which the wind can carry over the area to be seeded, such as larches, firs, spruces, most pines, etc.

The cuttings are made as much as possible in an oblong shape, with the longest side at right angles to the direction of the prevailing winds. The breadth of the clearing, on which occasional reserves of not too spreading crowns may be left, depends of course on the distance to which the wind can easily carry the seed which is to cover the cleared area. Observation and experience will determine the distance. In Germany, for spruce and pine, this has been found to be twice the height of the tree; for larch, five or six times the height; for fir, not more than one shaft's length. From 200 to 360 feet is perhaps the range over which seeding may be thus expected. One year rarely suffices to cover the cleared area with young growth, and it takes longer in proportion to the breadth of the cutting. This method is very much less certain in its forestal results than the next named, and more often requires the helping hand of the planter to fill out bare places left uncovered by the natural seeding. But it is the one that seems to interfere least with our present habits of lumbering, and with it eventually the first elements of forestry may be introduced into lumbering operations.

To be sure, it requires from three to eight times the area usually brought under operation, but instead of going over the whole area every year it may be operated in a number of small camps systematically placed along a central road connecting the different camps or cuttings with the mill.

As a rule the pine forests in Germany are reproduced by artificial plantations, the spruce forests by either natural or artificial regeneration, or both combined, while the beech forests are entirely reproduced as described above, oaks and other hard woods being usually planted, although a return to a more extended use of natural reproduction is noticeable.

IMPROVEMENT CUTTINGS—THINNINGS.

The principles which underlie the practice of thinning out young growths in order to accelerate their development have been theoretically well developed, but the practice in Germany remains behind the theory. The difficulty of disposing of the material taken out in the thinnings discourages the practitioner, and the financial value of the operation in the acceleration of the remaining crop is not fully appreciated.

A few results of German practice in thinning may serve to give an indication of its value.

A natural growth of pine (Scotch) which was thinned when six years old showed an increased rate of accretion three times as great as that of the part not thinned, which was also deficient in height growth.

A 50-year-old spruce (Norway) growth, having been twice thinned, showed an average accretion 22 per cent greater than the part not thinned.

A growth of spruce (natural sowing), slightly mixed with maple, aspen, willow, and ironwood, when 15 years old was opened to the poor population to take out firewood; thus one-half of the growth for a few years was thinned out irregularly. The part thus thinned eighteen years later contained four and one-half times more wood than the undisturbed part; the former contained trees of from 1 to 9 inches in diameter and 15 to 65 feet in height; the latter did not produce any above 5 inches in diameter and 48 feet in height.

Another experiment, made upon a pine growth 50 years old, showed that by interlucation the rate of growth within eleven years stood three to one and three fourths in favor of the thinned part.

Another writer planted Scotch pine 6 feet apart; two years later he planted the same ground to bring the stand to 3 feet apart; he thinned when fifteen years old, and carefully measured contents when twenty years old. Although the plantation was stocked on poor soil, yet the average annual accretion was found to be 2.13 cords (Austrian) per acre, a yield "which is unexcelled." The writer adds that "if in such growths the number of trees is reduced in the fifteenth to twentieth years to 280 trees per acre, the yield in sixty years might equal that obtained in one hundred or one hundred and fifty years in the old manner."

A plantation of Norway spruce, made with seed, was when thirty-three years old still so dense that it was impenetrable; hardly any increase was noticeable and the trees were covered with lichens. When thirty-five years old it was thinned, and again, when forty-two years old the condition of the growth was such as to make a thinning appear desirable; between the two thinnings, within seven years, the accretion had increased by 160 per cent, or 27 per cent yearly in the average, and the appearance of the trees had changed for the better.

A coppice of tanbark oak was thinned when fifteen years old on half the area; when twenty years old both parts were cut, and it was found that the thinned part yielded more wood and more and better bark than the unthinned part, and yielded in money 11.5 per cent more, although no higher price was asked for the better bark.

An area of 12 acres was planted, one-half with 2-year-old pine seedlings from the forest, the other half with seed.

Three thinnings were made with the following yield of round firewood (cut to billet length and over 2¼ inches in diameter) and brushwood (less than 2¼ inches in diameter).

The planted part yielded at the thinnings:

When —	Firewood.	Brush.
	Cords.	Cords.
10 years old	1.4	1.4
15 years old	4.9	2.8
18 years old	4.5	2.8
Total	10.8	7

The sowing was first thinned when 8 years old, yielding:

When	Firewood.	Brush.
	Cords.	Cords.
8 years old		2.8
10 years old		3.6
20 years old	3.2	1.4
Total	3.2	7.8

In twenty-four years the total yield, inclusive of thinning, was:

	Cubic feet of solid wood.
Planted part	3,405
Sowed part	1,908
In favor of planted part	1,497

Thinnings are usually made for the following purposes:

(1) Improvement cuttings, to improve the composition of the forest and give advantage to the better kinds.

(2) Interlocations, to improve the form and hasten development of young timber.

(3) Regeneration cuttings, to produce favorable conditions for seed formation and reproduction of the forest.

(4) Accretion cuttings, to improve rate of diameter growth in older timber.

Thinnings are to open the crown-cover, giving access to light and air, their object being to accelerate decomposition of the litter and turn it into available plant food; to improve the form and hasten the development of the remaining growth. The degree of thinning depends on soil, species, and age, and is best determined as a proportion between the present growth and that which is to remain with reference either to crown-cover, mass, or diameter.

Since it is observed that in the struggle for existence among the individual trees there are quite early some trees getting the advantage and becoming dominant, it is inferred that thinnings are most effective in the earlier period of the crop.

In discussing the degree to which the thinning is to be made, a classification of the trees according to the character of their development is made by German foresters as follows:

Dominant or superior growth.
- Class 1.—Predominant trees with highly developed crowns.
- Class 2.—Codominant trees with tolerably well developed crowns.
- Class 3.—Subdominant trees with normal crowns, but poorly developed and crowded above.

Dominated or inferior growth.
- Class 4.—Dominated trees with crowns poorly developed and crowded laterally.
  - (a) Crowns wedged in laterally, yet not overtopped.
  - (b) Crowns compressed, partly overtopped.
- Class 5.—Suppressed trees, entirely overtopped.
  - (a) Crowns still having vitality (shade enduring species).
  - (b) Crowns dying or dead.

The following illustration of the appearance of these tree classes will be found serviceable in understanding these relations.

Fig. 26. Tree classes. Classification according to crown development. Schematic. Class I (predominant) Nos. 1, 3, 6, 11, 16, 20, class 2 (codominant): Nos. 8, 13, 18; class 3 (subdominant): Nos. 9, 14, 17; class 4 (oppressed): Nos. 5, 7, 12; class 5 (suppressed, a): Nos. 2, 19; class 5 (suppressed, b): Nos. 4, 10, 15.

The degrees of thinning usually resorted to are the following:

(1) Slight thinning takes out trees of class 5.

(2) Moderate thinning takes out trees of class 5 and 4b.

(3) Severe thinning takes out trees of class 5, 4, and sometimes 3.

The time when the first thinning should take place is generally determined by the possibility of marketing the extracted material at a price which will cover at least the expense of the operation. This is, however, not always possible, and the consideration of the increase in value of the remaining growth, or rather of the detriment to the same by omission of timely thinning, may then be conclusive.

On good soil and on mild exposures interlucation may take place earliest, because here the growth is rankest and a difference in the development of the different stems is soonest noticeable.

Light-needing and quicker-growing kinds show similar conditions to those grown on good soil, and here, therefore, early thinnings are desirable. In these cases the thinnings have also to be repeated oftenest, especially during the period of prevalent height accretion. Absolute rules as to the time for interlucations and their periodical repetition evidently can not be given. The peculiar conditions of each individual case alone can determine this. The golden rule, however, is early, often, moderately. The right time for the beginning of these regular and periodical interlucations is generally considered to have arrived when the natural thinning out before mentioned commences and shows the need of the operation. This occurs generally when the crop has attained the size of hop poles. At this stage the well-marked difference in size of the suppressed trees will point them out as having to fall, and there will not be much risk of making any gross mistakes. Until the trees have attained their full height the thinning should remain moderate. From this time forward it will prove expedient to open out the stock more freely without ever going so far as to thin severely. Within the last few years new and revolutionary ideas regarding principles and methods to prevail in thinnings are gaining ground, which we have not space here to discuss.

UNDER-PLANTING.

All these manipulations experience modifications according to circumstances, different species and soil conditions requiring different treatment. One of the most interesting modifications, the results of which in a given district were fully exhibited, is the v. Seebach management in beech forests. Such a management, which contemplates the production of heavier timber in the shortest time, tries to take advantage of the increase in accretion due to an increase of light which is secured by severe thinning, and in order to prevent the drying out of the soil by such severe thinning a cover of some shady kind is established by sowing or planting. This cover gradually dies off under the shade of the old timber, the crowns closing again after a number of years. The rate of growth in a stand of 70 to 80 years was thereby increased from 51 cubic feet per acre and year to 77 cubic feet per acre and year, while a neighboring stand, otherwise the same but not so treated, increased by only 60 cubic feet, distributed over a larger number of trees.

The same method is applied to the production of heavy oak timber. In this case the oak growth is thinned out when about 60 years old and "underplanted" with beech. It may also be applied to older growths with advantage, as appears from the following results:

A stand of oaks 150 to 160 years old in 1846 was thinned to 96 trees per acre, averaging 37 cubic feet of wood per tree, the cleared space being "underplanted" with beech and spruce. In 1887 the oaks, now 190 to 200 years old, of which 57 trees only were left, contained 56 cubic feet in the average, thus growing during the last forty years more than one-half as much as during the one hundred and fifty to one hundred and sixty years previous to the operation, i. e., doubling the rate of growth. In this case, under the light-foliaged oaks, some of the beech and spruce developed sufficiently to furnish marketable material.

With Scotch pine it has been found in one case that while the average accretion of a stand 120 years old under ordinary condition was about 59 cubic feet per acre and year—the yield by thinning included—a stand underplanted with beech showed an accretion of 100 cubic feet per acre and year, besides much better log sizes and earlier supply of saw timber.

Translated into money an example from Bavaria may be cited as follows:

On 1 acre of pine 80 years old, underplanted at a cost of $2.85 per acre with beech now 10 years old, there were found—

	Yield of wood.	Average annual accretion per acre.
	Cubic ft.	Cubic ft.
105 pine	322	40
2,300 beech	156	39
Total	478	79

Supposing this stand to be left forty years longer, it may be figured that the pine would bring $650 and the beech $120; total per acre, $770, of which $19 was yielded in thinnings. White pine without undergrowings is expected to produce only $520 per acre when 120 years old.

FORESTERS, FORESTRY EDUCATION, AND FORESTRY LITERATURE.

To be sure, the highly elaborate system of forest administration and forest management here outlined could not be developed or maintained without a special high-grade education of those who direct the work. This education is provided for in the most ample manner, and consists not only in theoretical studies at schools, academies, and universities, but also in practical studies in the forest itself under the guidance of competent and experienced forest managers.

The course which applicants for positions in the higher administrative forestry service are expected to follow, with more or less modification in the different states, may be briefly outlined here:

After promotion from college the student goes into the woods for a short period (one-half to one year) to acquaint himself, under the guidance of a district manager, with the general features of the business he proposes to engage in, and thereby tests his probable fitness for it. He then visits for two and one-half or three years a forestry school (called academy when by itself, when at a university it is connected with the "faculty" for national economy), where theoretical studies with demonstrations in the forest are pursued.

After examination and promotion the applicant is bound at his own expense to occupy himself for two years at least in studying the practice in various districts, changing from place to place. If occupation can be found for him he is employed at small daily wages on some scientific or administrative work, always keeping an official diary of his doings and observations, certified to by the district manager with whom he stays, and which forms part of his final examination. For nine months during this time he must continuously perform all the duties of a lower official—a ranger—for a whole or part of a range, and sometimes also for a given time certain functions of a district manager. Then, after two years of law studies at a university, he enters into a close and difficult examination for a position as district manager, lasting eight to ten days. By passing this he is placed on the list of eligibles, and has thereby secured a right, enforcible in the courts if need be, to a position when a vacancy arises and his name is reached in the order of the list. This, in Prussia, may now be within eight or ten years after listing. During the interval he may be, and mostly is, employed on daily wages in various sorts of scientific and administrative work, such as revising and making new valuations, laying out roads, acting as tutor at the academies or as assistant to district managers, or else taking the place of a manager temporarily, etc.

The higher administrative offices are filled by selections from the managers, length of service counting only when special fitness for the kind of work required accompanies it; so that, as in the army, the highest officer has been through all the grades below, and is conversant with every detail of the service. The pay is small, graded in each kind of position according to length of service and somewhat according to the cost of living in different places. The honor of the position, to which usually other honors are added, its permanency, and the assurance of a pension, graded according to length of service, in case of disability or age, make up for small salaries. The salaries, subject to change from time to time, without adding the value of perquisites like houses, farm lands, etc., range about as follows in Prussia:

1 director (Oberlandforstmeister)	$3,600
4 forest councilors (Landforstmeister)	$1,800 to 2,400
33 chief inspectors (Oberforstmeister) (with additions for house and traveling up to $1,100)	1,050 1,500
89 inspectors (Forstmeister) (with additions for house and traveling up to $1,100)	900 1,500
679 district managers (Oberfoerster) (with additions up to $825 and house and field)	500 900
3,390 rangers (Foerster) (with house and additions up to $110)	260 360
319 guards (Waldwaerter)	100 200

The rangers (Foerster) follow different courses of instruction, part of which they receive in subordinate positions under district managers; while serving in the army in special battalions (chasseurs) they receive also theoretical instruction, which is supplemented in special schools. When finally promoted to the responsible position of rangers, in which much discretion and latitude are given them, their pay amounts to from $260 to $360, with a house and field, with the assurance of pension on withdrawal.

The following schools are provided for the higher grades of foresters:

Higher forestry schools in Germany for the education of forest managers.

[Austria and Switzerland included.]

Name of place.	State.	When founded.	Length of course (years).	Instructors of forestry branches proper.	Total number of instructors.	Average attendance of forestry students.
At universities:						
Giessen	Hesse	1825	3	3	(a)	40 50
Tübingen	Wurttemberg	1818	(b)	3	(a)	50 60
Munich	Bavaria	1878	(b)	8	a 18	c 90-100
At polytechnicums:						
Karlsruhe	Baden	1832	3	2	19	15 30
Zurich	Switzerland	1855	3	3	c 20	15 30
Vienna	Austria	1875	3	6	47	130-149
Separate academies:						
Aschaffenburg	Bavaria	1807	2	2	9	90 100
Tharandt	Saxony	1811	2½	3	10	100-125
Eisenach	Saxe Weimar	1830	2	3	8	65 75
Eberswalde	Prussia	1831	2½	6	14	100-150
Münden	do	1868	2½	5	13	40 60

a The entire corps of professors of the university. In Munich 18 professors are engaged in lecturing on subjects which concern forestry students; in Zurich, 20 professors. In Munich all studies can be followed in any year, as the students may select. The attendance varies, of course, widely in different years having been as high as 216 in Eberswalde and 129 in Münden. The above figures are for 1885 86.

b Not prescribed.

c During the winter of 1895 there were 110 students at Munich out of 527 forestry students at all forestry schools.

The following table will serve to give an idea of what instruction is to be had at these institutions:

Plan of studies at Forest Academy Eberswalde.

Subjects of instruction.	Whole number of hours.
FUNDAMENTAL SCIENCES.	
Natural sciences.	
General and theoretic chemistry	32
Special inorganic and organic chemistry applied	80
Physics and meteorology	80
Mineralogy and geognosy	60
Definition of minerals and rocks	20
Reviews for organic natural science	16
Botany in general and forest botany in particular	64
Anatomy of plants, vegetable physiology and pathology	60
Microscopy	20
Botanical reviews	20
Botanical excursions, each 2½ hours	80
General zoology	16
Vertebrates	80
Invertebrates, with special reference to forest insects	80
Zoological preparations	16
Zoological reviews	20
Zoological excursions, each 5 hours	96
Total natural sciences	840
Mathematics	
Geodesy	72
Interest and rent account	20
Wood measuring	20
Mathematical reviews and exercises	56
Surveying and leveling exercises, each 4 hours	192
Plan-drawing exercises, 2½ hours	80
Total mathematics	440
Economic sciences.	
Public economy and finances	48
Total sum of hours for fundamental sciences	1,328

Subjects of instruction.	Whole number of hours.
PRINCIPAL SCIENCES.	
Cultivation of forests	80
Forest implements	20
Geographical forest botany	48
Protection of forests	32
Forest usufruct and technology	80
Forest surveying	20
Appraising forests	80
Calculation of the value of forests and forest statistics	32
Administration of forest and hunting	48
Redemption of rights of usage	32
Forest history	40
Forest statistics	20
Review of various forest matters	56
Examinations	40
Forest excursions, each 4 hours	352
Total	980
SECONDARY SCIENCES.	
Jurisprudence.	
Civil law	72
Criminal law	32
Civil and criminal lawsuits and constitutional rights	40
Jurisprudence	36
Total	180
Construction of roads	32
Hunting	32
Shooting exercises, 2 hours each	96
Total sum of hours for secondary sciences	340
Grand total	2,648

	Per cent.
Fundamental sciences	50
Principal sciences	37
Secondary sciences	13

Average per instruction week (21 weeks in winter, 17 during summer; 2 winter courses, 3 summer courses):

$$\frac{2648}{93} = 28.5 \text{ hours, or per day, 4.9 hours.}$$

If we were to codify into a system the science of forestry as developed in Germany we might come to the following scheme, which exhibits the various branches in which a well-educated forester must be versed:

SYSTEM OF FORESTRY KNOWLEDGE.

I. FOREST POLICY—ECONOMIC BASIS OF FORESTRY (THE CONDITION).

Aspects.

1. *Forestry statistics.* (Areas, forest conditions; products. By-products: Trade; supply and demand; prices; substitutes.)
2. *Forestry economics.*
   (a. Study of relation of forests on climate, soil, water, health, ethics, etc.
   b. Study of commercial peculiarities and position of forests, and forestry in political economy.)
3. *History of forestry.*

Application.

4. *Forestry politics.* (Formulation of rights and duties of the State and of its methods in developing forestry; legislation, State forest administration, education.)

II. FOREST PRODUCTION—TECHNICAL BASIS OF FORESTRY (THE CROP).

Aspects.

5. *Forest botany.* (Systematic botany of arborescent flora; forest geography; plant and climate; biology of trees in their individual and aggregate life; forest weeds.
6. *Soil physics and soil chemistry with special reference to forest growth.*
7. *Timber physics.* (Anatomy of woods; chemical physiology and physical properties of woods. Influences determining same; diseases and faults.)
8. *Technology.* (Application of wood in the arts; requirements and behavior; mechanical and working properties; durability; special needs of consumers; use of by-products, waste materials, minor forest products.)

Application.

9. *Silviculture.* (Methods of growing the crop.)
   a. Natural reforestation; cutting for reproduction.
   b. Artificial afforestation; procurement of plant material; nursery practice, choice of plant material, methods of soil preparation, of forest planting.
   c. Improving and accelerating the crop. Cultivation, filling, thinning, pruning, undergrowing.
   d. Systems of management. Timber forest, standard coppice, coppice, etc.
10. *Forest protection.* (Against insects, climatic injuries, fire, cattle, etc.)
11. *Forest improvement and engineering.* (Treatment of denuded mountain slopes, shifting sands, barrens, swamp and moors, road building, etc.)
12. *Forest utilization.* (Methods of harvesting, transporting, preparation for market.)

III. FOREST ORGANIZATION—ADMINISTRATIVE AND FINANCIAL BASIS (THE REVENUE).

Aspects.

13. *Forest survey.* Ascertaining area and condition of the forest; ascertaining rate of accretion, yield.
14. *Forest valuation and statics.* Ascertaining money value of forest soil and forest growth as capital of the management and comparing financial results of various kinds of management.

Application.

15. *Forest regulation.* Establishing units of management and administration; determining working plans, distributing yearly or periodical cut, etc.
16. *Forest administration.* Routine methods, business practice, personnel, organization of service and mechanical operations.

LITERATURE.

In addition to the live teachings, which an able corps of professors impart at these institutions and that which competent managers are ready to impart to the young students in the forest itself, a large number of weekly, monthly, quarterly, and annual journals and publications are keeping the foresters and forestry students *au courant* with the progress of forestry science and forestry technique. Adding the publications of this nature which appear in Austria and Switzerland in the German language, and which have their constituency in Germany as well, we can make the

H. Doc. 181——17

following respectable list, not counting the journals of the lumber trade and other related publications. Those marked with an asterisk (*) are to be found in the library of the Division of Forestry; those marked (†) are considered the best or are most comprehensive; those marked (?) have been discontinued.

German forestry periodicals.

Name of publication.	Published at—	Issued	Established.
Allgemeine Forst u.Jagdzeitung * †	Frankfort on the Main	Monthly	1824
Aus dem Walde	Hanover	Irregularly	1865
Aus dem Walde	Frankfort on the Main	Weekly	(?)
Deutsche Forst-u. Jagdzeitung	do	Semimonthly	(?)
Forstliche Blaetter	Berlin	Monthly	1863
Forstlich-naturwissenschaftliche Zeitschrift * †	Munich	do	1892
Forstwissenschaftliches Centralblatt * †	Berlin	do	1856
Jahresbericht des schlesischen Forstvereins	Breslau	Annually	1811
Jahresbericht der preussischen F. u. J. Gesetzgebung	Berlin	do	1868
Land u. Forst wirthschaftliche Zeitschrift	Vienna	Quarterly	1886
Muendener forstliche Hefte*	Berlin	Irregularly	1892
Oesterreichische Forst zeitung *	Vienna	Weekly	1882
Der praktische Forstwirt fuer die Schweiz	Davos	(?)	(?)
Schweizer Zeitschrift fuer Forstwesen	Zürich	Quarterly	(?)
Tharandter forstliches Jahrbuch *	Dresden	Annually	1850
Verhandlungen der Forstvereine	Various	do	
Bericht ueber die Versammlung deutscher Forstmaenner	do	do	
Zeitschrift fuer Forst-u. Jagdwesen * †	Berlin	Monthly	1869
Zentralblatt fuer das gesammte Forstwesen * †	Vienna	do	1875
Zeitschrift der deutschen Forstbeamten	(?)	(?)	(?)

Should the reader wish to collect a library of the most modern thought on any or all subjects pertaining to forestry in Germany the list of books contained in the library of the Department of Agriculture, a catalogue of which has been published, with over 1,200 numbers and probably 2,000 volumes, would give him a good selection.

FORESTRY ASSOCIATIONS.

Forestry associations thrive better in Germany than in the United States and are of a different character: they are associations of foresters, who practice what they preach. There is no more need of a propaganda for forestry than there would be here for agriculture, and the discussions, therefore, are moving in technical, scientific, and economic directions. Besides some thirty or forty larger and smaller local associations, there is held every year a forestry congress, at which the leading foresters discuss important questions of the day.

FOREST EXPERIMENT STATIONS.

In addition to all these means of education and of advancement of forestry science, and in addition to the demonstration forests connected with the various schools of forestry, there has been developed in the last twenty years a new and most important factor in the shape of forest experiment stations, which are also mostly connected with the forestry schools. If forestry had a strong and well-supported constituency before, this additional force has imparted new impulses in every direction.

The first incentive for the establishment of these stations came from the recognition that the study of forest influences upon climate could be carried on only with the aid of long-continued observations at certain stations. Accordingly, during the years 1862 to 1867, forest meteorological stations were instituted in Bavaria, which, under the efficient direction of the well-known and eminent Dr. Ebermayer, for the first time attempted to solve these and other climatic questions on a scientific basis. The results of these and other observations have been fully discussed in Bulletin 7 of the Forestry Division and are briefly recorded in this report.

While these stations were continued and others added in all parts of the country, an enlargement of the programme was soon discussed with great vigor, leading (between the years 1870–1876) to the institution of fully organized experiment stations in Prussia, Bavaria, Saxony, Thuringia, Wurtemberg, Baden, Switzerland and Austria following in the same direction; all of these finally combining into an "association of German forest experiment stations," similar to the association of agricultural experiment stations in our country. Thus the science of forestry, which hitherto had been developed empirically, has been placed upon the basis of exact scientific investigation, the fruit of which is just beginning to ripen in many branches.

We in the United States are fortunate, in that we can learn from the experience and profit from the assidnous work of these careful investigators. While we may never adopt the admirable administrative methods that fit the economic, social, and political conditions of Germany, we shall ever follow them where the recognition and utilization of natural laws lead to the practical acknowledgment of general principles and to desired economic results in forest culture.

FOREST MANAGEMENT IN BRITISH INDIA.

In order to show how the transfer of German methods may work advantageously, even in a country entirely differently conditioned, the results obtained by the forest management in British India are here briefly stated.

India, with a total area of nearly 1,500,000 square miles or 936,000,000 acres (an area about one-half that of the United States without Alaska), has a population of about 270,000,000, or four times as great as that of the United States.

Of the entire area about 950,000 square miles, or 63 per cent, are under British rule, the remaining 550,000 square miles, with a population of about 53,000,000, being divided among a large number of more or less independent native States.

Of the entire population about 70 per cent are farmers and farm laborers, who cultivate about 200,000,000 acres of land, 30,000,000 of which is irrigated. The greater part of the main peninsula is a high plateau with steep descents to the ocean, both on the western and eastern coast.

To the north of this plateau is a broad, fertile, river plain extending from the upper Bramahputra to the mouth of the Indus, a distance of nearly 2,000 miles, without rising more than 900 feet above sea level. North of this large and densely settled Indo-Gangetic plain, and forming the barrier between India and Thibet, is the great Himalaya Mountain system, drained by the three great river systems of northern India.

More than half of India lies within the Tropics and over 90 per cent is farther south than New Orleans, the latitude of which is 30°. From this it is apparent that the climate is generally hot, but, owing to diversity of elevation and peculiarities of the distribution of rainfall, it is by no means uniform.

The rains of India depend on extraordinary sea winds, or "monsoons." and their distribution is regulated by the topography of land and the relative position of any districts with regard to the mountains and the vapor-laden air currents. Thus excessive rainfall characterizes the coast line along the Arabian Sea to about latitude 20° N., and still more the coast of Lower Burmah, and to a lesser extent also the delta of the Ganges and the southern slope of the Himalayas. A moderately humid climate, if gauged by annual rainfall, prevails over the plateau occupying the large peninsula and the Lower Ganges Valley, while a rainfall of less than 15 inches occurs over the arid regions of the Lower Indus. In keeping with this great diversity of climate, both as to temperature and humidity, there is great variation in the character and development of the forest cover. The natural differences in this forest cover are emphasized by the action of man, who for many centuries has waged war against the forest, clearing it permanently or temporarily for agricultural purposes or else merely burning it over to improve grazing facilities or for purposes of the chase. Thus only about 25 per cent of the entire area of India is covered by woods, not over 20 per cent being under cultivation, leaving about 55 per cent either natural desert, waste, or grazing lands. The great forests of India are in Burmah; extensive woods clothe the foothills of the Himalayas and are scattered in smaller bodies throughout the more humid portions of the country, while the dry northwestern territories are practically treeless wastes. In this way large areas of densely settled districts are so completely void of forest that millions of people regularly burn cow dung as fuel, while equally large districts are still impenetrable, wild woods, where, for want of market, it hardly pays to cut even the best of timbers.

The great mass of forests of India are stocked with hardwoods (i. e., not conifers), which in these tropical countries are largely evergreens, or nearly so, and only a small portion of the forest area is covered by conifers, both pine and cedar, these pine forests being generally restricted to higher altitudes. The hardwoods, most of which in India truly deserve this name, belong to a great variety of plant families, some of the most important being the Leguminosæ, Verbenaceæ, Dipterocarpeæ, Combretaceæ, Rubiaceæ, Ebenaceæ, Euphorbiaceæ, Myrtaceæ, and others, and but a relatively small portion of them represent the Cupuliferae and other important hardwood timber families so characteristic of our woods.

In the greater part of India the hardwood forest consists not of a few species, as with us, but is made up of a great variety of trees unlike in their habit, their growth, and their product, and if our hardwoods offer on this account considerable difficulties to profitable exploitation, the case is far more complicated in India. In addition to the large variety of timber trees there is a multitude of shrubs, twining and climbing plants, and in most forest districts also a dense undergrowth of giant grasses (bamboos), attaining a height of 30 to 120 feet. These bamboos, valuable as they are in many ways, prevent often for years the growth of any seedling tree, and thus form a serious obstacle to the regeneration of valuable timber. The growth of timber is generally quite rapid; the bamboos make large, useful stems in a single season. Teak grows into large-size saw timber in fifty to sixty years. But in spite of their rapid growth and the large areas now in forest capable of reforestation, India is not likely to—at least within reasonable time—raise more timber than it needs. In most parts of India the use of ordinary soft woods, such as pine, seems very restricted, for only durable woods, those resisting both fungi and insects (of which the white ants are specially destructive), can be employed in the more permanent structures, and are therefore acceptable in all Indian markets.

At present teak is the most important hardwood timber, while the deodar (a true cedar) is the most extensively used conifer. Teak occurs in all moist regions of India except the mountain countries, never makes forests by itself (pure forests), grows mixed with other kinds, single, or in clumps, is girdled two to three years before felling, is generally logged in a primitive way, commonly hewn in the woods and shipped—usually floated—as timber, round or hewn, and rarely sawn to size. Teak is as heavy and strong as good hickory, has little sapwood, stands well after seasoning, and is remarkably proof against decay and the still more dreaded white ants, and is really the only important export timber of India, about $2,500,000 worth having been shipped in 1894-95, bringing about $1 per cubic foot, or more than four times as much as good pine timber in the market.

As will be seen from the following figures timber forms only about 20 per cent of the export of forest products, which consist chiefly of lac, the basis of shellac (really the product of an insect) and of tanning materials:

Exports of forest products from India, 1894-95.

Lac (basis of shellac)	$7,000,000
Teak	2,800,000
Myrobalans	2,300,000
Cutch and gambier	1,450,000
Caoutchouc	550,000
Fancy woods—sandal, ebony, rosewood	290,000
Cardamoms	140,000
Total	14,530,000

The imports of timber into India have so far been very insignificant. Attempts at introducing American coniferous timber (pine, spruce, larch, and hemlock) from the Pacific coast have not been successful, though it would seem that some wood goods, such as boxes, sash and door, and cheap furniture, should find a favorable and extensive market if once the trade is established. Perhaps a treatment of these materials with some of the new fireproofing substances could be made to render them at the same time more resistant to white ants and other insect borers, and thus procure for them several important advantages at once.

In the past the people of India, as far as known, never realized the importance of their forests. They were cleared, destroyed, mutilated at all times and in all places, and the use of wood never seems to have formed an important factor in Hindoo civilization.

With the advent of foreign commerce the exploitation of the forests for the more valuable export timbers received a new stimulus and the forests were culled regardless of the future, either of forest or people. This matter was aggravated by the construction of railways, which, in themselves large consumers, also offered a premium on all that contributed to increased traffic. When, finally, it was noticed that the demands of timber for public works in some localities could

www.ingramcontent.com/pod-product-compliance
Lightning Source LLC
Chambersburg PA
CBHW022013190326
41519CB00010B/1501